Cardio-Thoracic Injury, Essentials All Critical Care Nurses Need To Know

Cardio-Thoracic Injury, Essentials All Critical Care Nurses Need To Know

Sameh Elhabashy
CCRN, MScN, BSc
Critical Care Nursing Department
Faculty of Nursing
Cairo University

2015

First Printing: 2015

ISBN: 978-1-329-75327-3

MOSBY
Elseveir Health

1600 John F. Kennedy Boulevard
Suite 1800
Philadelphia, PA 19103-2822
USA www.cu.edu.eg.com

Special discounts are available on quantity purchases by corporations, associations, educators, and others. For details, contact the publisher at the above listed address.

U.S. trade bookstores and wholesalers: Please contact Sameh Elhabashy Tel: (+20) 100-7653998; Fax: (+20) 23657190 or email: Sameh17@cu.edu.eg

Sameh Elhabashy

To my lovely wife and children

Thank you. Without your support and persistence, I would have never accomplished this work.

Content

Acknowledgements

I would like to express my great thanks and appreciation to my family, parents, colleagues, teachers and my students who are always willing to provide their support and guidance. I also appreciate the efforts of (MOSBY) acquisitions editors and there guidance.

Sameh Elhabashy

Preface

Chest trauma is a significant source of morbidity and mortality in the world especially for people in their productive ages which affecting the economy and social adherence of the communities. This book focuses on the nurses' management of chest traumatic patients based on theoretical, ethical, legal and comprehensive practical background. Whether you are a beginner or an experienced clinician, I hope that you find this book enjoyable and clinically relevant that because of simple presentation of ideas and supported figures.

Cardio-Thoracic Injury, Essentials All Critical Care Nurses Need To Know

Introduction:

Cardio-Thoracic Injury is one of the leading causes of morbidity, mortality, and Life-threatening complications. It ranks third behind head and extremity trauma in major accidents in developing countries. This book developed to be a much-needed reference for nurses practicing this challenging field. This book is novel in its approach to chest emergency topics, it describes simply the best and most current methods to care for patients with thoracic injury including; theoretical frame, ethical and legal consideration, initial assessment, path-physiology, generation of differential diagnoses, problem solving, general management in addition to management of challenging conditions based on presenting symptoms. Unlike other textbooks, in which the patients' diagnosis is known, this book approaches is providing the initial management of clinical problems without full awareness of the final diagnosis. The book is advised by clinicians, educators, and researchers in the field of emergency nursing after wide search the most updated evidence based practices. Whether you are a beginner or an experienced nurse, this book will be useful because of simple presentation of ideas and supported figures.

Magnitude of Chest Trauma

Trauma is the leading cause of mortality and disability, especially during the productive age, and is the third most common cause of death. Accidents which are unexpected and unplanned events are becoming the major epidemic of the present century. Approximately one quarter of civilian trauma deaths are caused by thoracic trauma and many of these deaths can be prevented by prompt diagnosis and correct management. In spite of the high mortality rates, about 90% of the patients with life-threatening thoracic injuries can be managed by a simple intervention, e.g.; drainage of the pleural space by tube thoracostomy, Dalal, Nityasha, & Dahiya (2009).

Chest injuries can be sustained in isolation or in association with multiple injuries. Life-threatening complications may ensue because organs that are vital to survival of the organism are situated within the thoracic cavity. These complications include airway obstruction, tension pneumothorax, wide open pneumothorax, flail chest, cardiac tamponade and massive hemothorax. Severe multiple trauma is of extraordinary medical and social and economical importance.

Epidemiology of Chest Trauma

A few data were found to describe incidence of chest trauma. One of these reference is "Demirhan, Onan & Halezeroglu, (2009)" whom Conduct a Comprehensive, retrospective, and analytical study of 4205 patients with trauma at trauma hospital in Turkey within 10-year and reviled the following table.

(Table 1): Percentage of Various Injuries at 4205 Injured Patients Within 10-Year.

Type of traumatic injury	Percentage
Extremities	33%
Head and neck	30%
Chest	24%
Abdomen	13%
Total	**100 %**

In this mentioned study the morbidity rate in all victims was 25.2%. The mortality rate was 9.3% for all patients and was 6.8% in blunt, 1.4% in penetrating, and 17.7% in associated organ injuries. Mortality and Injury Severity Score (ISS) increased in patients having early surgery (P=0.001). Although most patients could be managed with conservative approaches, early thoracotomy was required in some cases. We believe that urgent hospital admission, early diagnosis, and multidisciplinary approach are very important to improve outcome.

Chest injuries are seen with increasing frequency in urban hospitals. The profile of chest injuries depends on the size of the hospital and the level of trauma center. The data regarding the true incidence of chest trauma are scant, Kulshrestha, Munshi & Wait, (2004). Chest trauma ranks third behind head and extremity trauma in major accidents in the United States. The motor vehicle accident is the most common etiology (70 %). Within the thorax, the chest wall itself is the most often injured. Many of these injuries are of moderate severity and rarely require surgical intervention. The majority of chest trauma requires careful surveillance to identify those patients who require operative correction. Improvement in vehicle safety, moderation of speed, and continued education should reduce the incidence and severity of chest trauma, LoCicero & Mattox (1989).

The first cause of death in Egypt is trauma or injury it ends life for nearly 30 000 persons every year it approximately 20 % of total deaths in Egypt in addition to 350 000 hospitalized patients every year. Moreover 8.2 % injured patients dead due to chest trauma (figure 1) it consider the second cause of injury after head injury , many families thus loss a son or father, which affects psychosocial and economic growth and development of Egyptian community that because the highest number of deaths. Many families thus lose a son or father, which affects the psychosocial and economic growth and development of Egyptian community that because the highest number of deaths or injuries occurs among men in the age group 15– 44 years (WHO, 2012).

5

Figure (1): Percentile Distribution of Death Causes According to Type of Injury among Traumatic Patients.

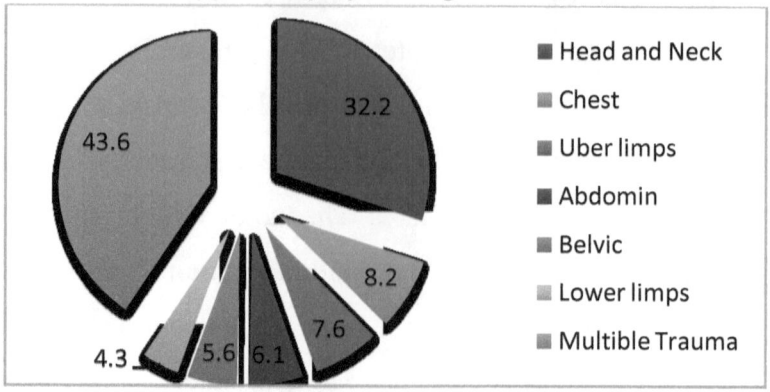

Theoretical Aspect of the Emergency Care for Acute Traumatic Patients:

Theories are patterns that guide the thinking about, being, and doing of nursing. Nursing theories address the phenomena of interest to nursing, including the focus of nursing; the person, group, or population nursed; the nurse; the relationship of nurse and nursed; and the hoped-for goal or purposes of nursing. Based on strongly held values and beliefs about nursing, and within contexts of various worldviews, theories are patterns that guide the thinking about, being, and doing of nursing. They provide structure for developing, evaluating, and using nursing scholarship and for extending and refining nursing knowledge through research. Nursing theories either implicitly or explicitly direct all avenues of nursing, including nursing education and administration.

Nursing theories provide concepts and designs that define the place of nursing in health and illness care. Through theories, nurses are offered perspectives for relating with professionals from other disciplines that join with nurses to provide human services. Nursing has great expectations of its theories. Theories must, at the same time, provide structure and substance to ground the practice and scholarship of nursing and also be flexible and dynamic to keep pace with the growth and changes in the discipline and practice of nursing.

Application of Nursing Theories to triage and Decision-Making:

Theory in nursing is frequently thought of as being a mainly academic exercise with little relevance to the everyday practice of nursing. Triage nursing have been investigated by several nurse researchers, however, most have not clearly articulated a theoretical or conceptual framework. Many nursing theories do not fit this description and should therefore, in the opinion of the authors, be thought of as models. The recognition primed decision (RPD) model is based on research about Decision Making (DM) under uncertain conditions such as time pressure, limited time available, high stakes, and changing cues. The context of emergency triage nursing (DM) is congruent with the (RPD) model. The authors propose that the (RPD) model can serve as a foundation for research that seeks to understand (DM) by triage nurses with the aim of yielding new knowledge that is useful for their practice, Reay & Rankin, (2013).

Emergency nursing is a specialty in which the nurse cares for patients in the emergency or critical phase of their illness or injury, focusing on the level of severity and time-critical interventions. Whilst collaborating with members of the emergency team, the emergency nurse plays a crucial role in the identification and care of patients with medical, surgical and injury related emergencies. The emergency nurse identifies life threatening problems, prioritizes the care, carries out resuscitative measures with appropriate management and provides information and emotional support to the patient and his/her family within a supportive health care environment. There is, however, no current consensus on issues of minimum level of education and training, delineation of roles and responsibilities, and/or effective staffing patterns for emergency nurses in Africa to enhance optimal patient outcomes.

In November of 2011, an international emergency nursing workgroup was convened in Cape Town, South Africa, in conjunction with the Emergency Medicine Society of South Africa's biannual conference, and with the support of the African Federation for Emergency Medicine and the Emergency Nurses' Society of South Africa. This workshop was attended by emergency nurses (both academic and clinical) from Tanzania, South Africa, Botswana, Ethiopia, Namibia and the United States, who participated in discussions surrounding the development of a framework for emergency nursing practice in Africa. It was decided that this framework should delineate the levels of practice, criteria for these different levels, evaluation of cognitive and psychomotor skill sets, and movement between levels of emergency nursing practice in Africa. The resultant framework has implications for nursing education and training, continuing education, and staffing at both institutional and regional levels throughout the continent and possibly further afield

Benner's Emergency Nursing Framework:

Benner's (1984) framework provides a well-validated schema to use to determine the emergency nurses level of expertise, including the knowledge and skill required for each of the levels within the framework. The majority of participants in the workgroup was familiar with this framework and agreed with its applicability to a broad range of both practice and educational settings. The usefulness of this framework was immediately clear; both clinicians and academics felt that it could be used along the educational and practice continuums. The participants also felt that a clearer picture of the expertise of each level of emergency nurses would be available, which could be used to develop guidelines for appropriate staffing patterns in a given care setting or across a region, and also provide guidelines for clinical education within emergency nursing practice Benner describes five levels of skill acquisition and development based on the Dreyfus model, positing that expertise is developed through skill acquisition. As a nurse passes through the five levels of development namely; Novice, Advanced Beginner, Competent, Proficient and Expert, there is a change in the perception of the job or skill base.

The novice level is when the nurse "has no background experience of the situation in which he/she is involved". Benner also points out that any nurse, even an expert, can find themselves at this level when changing specialties. However, as the nurse progresses through the five stages, he/she is able to incorporate more and more textbook knowledge and synthesize this into practical application to clinical situations. The time nurses spend in the novice and advanced beginner stages can vary from a few months to several years. The novice nurse is in orientation and gaining knowledge and experience in both clinical and technical arenas. The novice nurse works under the watchful guidance of a preceptor to collect objective information according to the guidelines and rules, and seeks assistance in making clinical decisions. The advanced beginner knows the rules, is guided by policies and follows them exactly and without deviation. He/she is unable to prioritize well, because he/she "treats all attributes and aspects as equally important because aspect recognition is dependent on prior experience." At this stage the nurse is governed by the tasks that need to be performed rather than patient responses or overall context. This stage requires strong preceptorship to teach the new nurse to determine the order of priority in her practice.

The competent nurse has been practicing long enough to acquire sufficient clinical experience to be able to manage a patient's care efficiently and appropriately. Competent nurses use standardized policies and procedures and routinely use hospital resources to solve problems. They are still, however, somewhat limited in their understanding of the holistic scope of the patient situation. Care is delivered based on a systematic approach guided by previous experience in recurrent situations. They are involved in the present situation, therefore their priorities are that of the moment and much may be missed.

The proficient nurse perceives the patient situation in a more holistic way and has in-depth knowledge of nursing practice. He/she can view the situation, drawing information from a variety of sources; lab work, patient presentation, patient concerns, etc., and hone in on the most salient aspects of the problem. There is much more involvement with patients and their families, and an increased confidence in personal skills and abilities. In this stage, says Benner, the "holistic understanding of the proficient nurse improves decision making".

Proficient nurses will learn best with case studies that allow them to view a situation from several angles. Instrumental to the progression from competent to proficient and expert nurses is what Benner calls the "paradigm case"; a patient situation which shifted the nurses expectations in such a way as to change his/her thinking. Benner states that such a case allows "the proficient clinician to compare past whole situations to current whole situations", and, as such, has a store of comparative situations to draw from in evaluating any current situation. The nurse begins to envision and create possibilities.

The expert nurse is the most skilled, according to Benner's framework, and no longer relies on analytical principles to guide his/her decision making. An expert nurse has enough clinical knowledge and paradigmatic situational experience to grasp, as a whole, the problem, its implications, and the priorities for treatment. Their intuitive skills arise from knowledge grounded in experience. Practice at this stage is characterized by a self-directed, confident and innovative approach to achieve the best possible outcome for the patient and family. This framework was empirically tested using qualitative methodologies and thirty one competencies, and seven domains of nursing practice were identified. Subsequent research suggests that the framework is useful and applicable in describing practice levels. The strength of the Benner model is that it is a data-based research that contributes to disciplinary knowledge as a philosophical theory of nursing.

The development of the framework for emergency nursing practice in Africa has clear implications for education. Guided by this framework for emergency nursing practice, undergraduate and postgraduate curriculums across the continent will need to be refined to encompass the suggested knowledge and skills for each level The goal in the development of this framework is to provide a continuum on which nurses can be moved towards an expert level of practice in order to provide quality care and optimal patient outcomes.

The pathway to accomplish the outcome is to keep a constant flow of emergency nurses at all of the experience levels specified by Benner, with an emphasis on regular advancement through consistent and well-defined practical and didactic emergency nursing education.

Initial pre-licensure education should take into account the need for expanded critical thinking and psychomotor skills in the absence of resources. Incorporation of these skills within clinical practice is relatively new in some parts of Africa, (2011), but these are essential components of nursing education, as supported by the World Health Organization standards for global nursing education. Research in the United States suggests that baccalaureate education is associated with better critical judgment skills and patient outcomes, and should be the entry point for nurses. Whilst this may be unrealistic in the African context, the continuing education for nurses however should focus on increasing the critical application skills of nurses, and not focus solely on content delivery.

A concerted effort should be made to develop a practice role for advance practice nurses (APRNs) in the area of emergency nursing. Given the relative dearth of physician providers, care gaps could be addressed with the use of APRNs. These emergency nurses should be theoretically trained to the 'proficient' or 'expert' level and should also be identified and recognized as practice leaders at the institutional or regional levels throughout Africa.

The framework will also guide appropriate clinical education to support growth in clinical judgment through the levels. For example, strong preceptorship should be included for novice and advanced beginner level emergency nurses to support recognition and evaluation of paradigm cases for the proficient nurse.

Mobility between levels the strength of the framework lies in the ability to evaluate emergency nurses' cognitive understanding of patient situations using the (Lasater Clinical Judgment Rubric) and thus place them within the levels as described by Benner. Use of these tools also gives educators and clinical preceptors a focus for guiding continuing education that will allow emergency nurses to improve their understanding and cognitive skills and move along the continuum of expertise. As emergency nurses gain experience and improve clinical judgment, they can demonstrate increased expertise and advance to a higher level of practice.

The use of this framework allows for the identification of expertise distribution among nurses in a given institution or geographic area. A proposed staffing pattern includes an 'expert' nurse at the hub, supervising up to four 'proficient' nurses, who in turn can supervise four 'competent' nurses. This staffing pattern can vary depending on the expected patient volume and acuity.

The framework has the potential to guide nation-specific healthcare human resources, allowing countries to develop efficient regional staffing that effectively supports nurses providing emergency care to make fast decisions and access higher level care.

The role and responsibility level of each nurse would be dependent on their level of expertise, For example, the nurses at the higher levels would be expected to impart their expert skills to nurses at the novice and advanced beginner levels, whilst also using their clinical judgment to oversee larger clinical scenarios that may include the highest acuity patients or managing the flow in the department. Novice nurses may need a closer supervision whilst functioning from protocol driven applications of patient care. Utilizing this grid would not only assist the nurse in his/her own professional development, but provide a framework for hospitals and other organizations to plan for safe and effective staffing and succession planning, Wolf, et al, (2012) .

Ethical Considerations in Emergency Nursing

Working in the emergency department (ED) gives rise to unique ethical considerations. Ethical problems are often exacerbated by time constraints, lack of detailed information, and a high incidence of impaired cognitive abilities in the patients. When patients arrive in the (ED), the triage nurse has little time to gather detailed information. Instead, a quick assessment is completed and actions are taken based on protocols, rather than the patient's preferences.

Emergency Nursing Code of Ethics:

The code of ethics directs nurses to maintain their high competence levels, exercise sound judgment in protecting the lives and privacy of patients and their families, and to practice with compassion, giving respect for human dignity, and respecting each individual for whom they are. The Emergency Nurses Association has published a Code of Ethics by which ED nurses are to guide their practice.

Ethical Expectations in (ED):

The expectation of services in the (ED) is to treat patients as well as inform them of their medical conditions. Overall, the goals of the (ED) staff are to quickly treat acute illnesses and injuries, minimize suffering and loss of functioning, and protect life. In the execution of the goals is the ethical principle of beneficence or the obligation of staff to improve the outcome for the patient.

When patients are unable to make decisions for them, the duty of the (ED) nurse is to advocate beneficence. This means to provide an objective view of what is best for the patient. This concept can become challenging when conflicts exist between clinically indicated treatment and the patient's religious or cultural values. An example of this is during active (CPR) when (ED) staff is performing an extensive variety of aggressive medical treatments without seeking approval or consent from any legal representative of the patient. The resuscitation efforts may conflict with the patients religious or cultural beliefs in regard to the receipt of blood products; however, this is in the best interest of the patient and is a reflection of beneficence.

Autonomy Begets Responsibility:

Autonomy brings with it an obligation to respect the choices of others, such as a patient's right to self-determination. It is the role of the (ED) nurse to ensure patients have accurate and comprehensive information to make informed decisions regarding treatment. Additionally, the nurse needs to ensure patients understand the potential benefits and success of certain procedures, but that outcomes are not guaranteed. In some instances, patients may need to be treated without informed consent if the intervention is essential to the preservation of life. However, when a treatment can result in serious harm, informed consent must be obtained.

The decisions people make may appear to be irrational, but if they are consistent with their internally-held beliefs, the nurse must advocate for them, as outlined by the American Nurses Association Code of Ethics."Nurses have the responsibility to promote health, and to advocate for the protection of the safety and self-demining rights of the patient," ANA says. This means the nurse must also advocate for terminally ill patients who choose to forgo life-sustaining treatment, as expressed verbally, in a living will, or a form of communication that is appropriately executed on the behalf of the patient.

Nurses must also act fairly to all persons regardless of gender, race, socioeconomic status, cultural background, or the ability to pay. All patients have the right to a standard of care as outlined in the Emergency Medical Treatment and Active Labor Act, which mandates access to quality emergency medical treatment to all whom seek it. The concept of non malficence, meanwhile, means to cause no harm, which is crucial to maintaining the integrity of staff and patient trust. (ED) nurses must ensure the safety of their patients in their care to the best of their ability. Additionally, it is the responsibility of the nurses to protect themselves, their coworkers and their patients against violent acts by known perpetrators, other patients or visitors. When violence is beyond that which can be handled by the nursing staff, then authorities must be brought in.

Nonmaleficence:

"Nonmaleficence" is defined as the duty to do no harm. This principle is the foundation of the medical profession's Hippocratic Oath; it is likewise critical to the nursing profession. Inherent in the Code of Ethics for Nurses (ANA, 2001), the nurse must not act in a manner that would intentionally harm the patient. Although this point appears straightforward, the nature of health care dictates that some therapeutic interventions carry risks of harm for the patient, but the treatment will eventually produce great good for the patient. Classic examples of this are chemotherapy and bone marrow or stem cell transplantation procedures. Both Interventions can make patients sicker for a time, posing a risk for complications such as opportunistic infections, but the possibility of achieving a cure or remission of disease may justify the temporary harm. The concept that justifies risking harm is referred to as the principle of "double effect."

Beneficence:

Beneficence is commonly defined as "the doing of good" and is one of the critical ethical principles in health care. In determining what is "good," nurses should always consider one's actions in the context of the patient's life and situation. Although this sounds simple, health care providers are challenged daily when what is good for the patient may also cause harm to the patient or is in conflict with what the patient wants. Suppose, for example, that an elderly patient has become confused, especially at night, and is at high risk for falls. She has fallen at home twice. Now that she is confused, she is at even more risk for a fall, especially as confusion can become worse at night in the elderly. The health care team decides that the patient needs a sitter to stay with her in her room all night.

The patient objects stridently, because she is very dignified and proud and enjoys her privacy. She complains that she "doesn't Need a babysitter" and cries for a long time. The Patient is prevented from falling but is psychologically distressed because of limitations on her independence, freedom, privacy, and dignity. Virtually everyone would agree that promoting good and avoiding harm are important to all human beings—and certainly to health care professionals. It may seem surprising, therefore, how often conflicts occur surrounding the principle of beneficence. A beneficent act may conflict with other ethical principles, most often autonomy. Even Though a nurse or physician may understand that a particular treatment has a benefit for the patient, the patient may decide to forego that treatment (autonomy) for a variety of reasons. In this instance, the health care provider should avoid acting paternalistically and recognize that the patient remains in a position of self-determination.

Justice:

The principle of justice is that equals should be treated the same and that unequal should be treated differently. In Other words, patients with the same diagnosis and health care needs should receive the same care, and those with greater or lesser needs should receive care that is appropriate to their needs.

Basic to the principle of justice are questions of who receives health care and whether health is a right or a privilege. These questions have been central to discussions surrounding health care reform in the United States in the past several years. Such questions involve the allocation of resources: how much of our national resources should be appropriated to health care; what health care problems should receive the most financial resources; what persons should have access to health care services? Numerous models have been developed for distributing Health care resources.

These Models provide suggestions for distribution have merit, but no single one is adequate in ensuring a just model for the distribution of health care resources. These include the following:

1. To each equally
2. To each according to merit
3. To each according to what can be acquired in the marketplace
4. To each according to need

Determining Decisional Capacity:

Like nurses, patients also have ethical obligations. They are expected to participate in their own care while collaborating and cooperating with ED staff, and should respect triage decisions and prioritization. Patients must always provide informed consent autonomously and voluntarily. This is achieved when patients are competent to agree or disagree to with proposed interventions and can sign consent their name. For those who are not competent or are unable to sign the consent forms, a surrogate will then assume responsibility. If a surrogate is not available, attempts will be made to contact one through acceptable modes of communication.

In some cases, patients may request that the (ED) physician make the decisions for them. Determining decisional capacity involves assessment of patients' ability to understand their and deliberate and articulate their healthcare preferences. Decisional capacity is dynamic and can improve or decline rapidly in an emergency setting. Diminished decisional capacity can vary during an emergency and can be reversible in certain states, such as intoxication, hypoxia, sedation and extreme stress. All efforts should be made to ensure the reversible cases are treated so the patient can make the most rational and autonomous choices. Decisional capacity is assessed in each (ED) patient, using indices such as ability to give a reasonable medical history, to cooperate with evaluation, and to understand the recommended treatments. Refusing to have a small laceration suture is one thing, but being unwilling to be admitted for treatment following a cardiac event is another.

Implied consent which based on the assumption that every rationale human being wants to live as long as possible (O'Neill, 2003). Consent is implied when it is impossible to be obtained immediately prior to performing a life-saving intervention or treatment (University of Washington School of Medicine, 2011). In emergency situations, staff should attempt to consult with the patient's attending physician or with another physician, and this transaction must be noted in the documentation McElroy, (2012).

Medical Futility:

When a person is unable to give consent, staff acts in accordance with the ethical concepts of "doing the greatest good" for each patient based on the implied consent or "What would I do if this were me?" while simultaneously applying the standard ethical principles of practice. The concept of medical futility is based on the notion of commonsense and acceptable levels of probability.

In essence, futile treatments are those that preserve permanent states of unconsciousness, or fail to end a patient's total dependence on intensive medical treatment (University of Washington School of Medicine, 2011). In the (ED), staff must be especially aware of this concept because of time constraints, or absence of relatives or ability to communicate with the patient. Some treatments are automatically carried out in the ED often before a detailed history has been obtained. Electronic records assist in these situations, but electronic records are not yet universal, lack of information continues to present difficulty. Additionally, consideration that future treatments may be futile is not basis to terminate all current treatments from being performed.

In some cases, supportive measures will be provided, and although certain treatments may be withheld, it is important to remember to maintain the support comfort measures as well as adequate communication for patients, family and friends Arras, (1993).

Refusal of Care:

Ethically patients have the right to refuse care. It is the responsibility of the ED nurse to ascertain if the patient has enough information to make an informed decision regarding refusal of care. If the reasoning for the decision is irrational, the nurse is responsible for ensuring that the patient has all necessary information. It is the responsibility of the nurse to make sure the patient has more information by which they can make a better informed decision, McElroy, (2012).

Ethical Problems

An ethical problem is one in which no clear answer exists for all. Nurses' previous experience and education can assist in the decision making process, and nurses should never hesitate to solicit input from colleagues when necessary. Although algorithms are available to guide care in cases of cardiac arrest and trauma, each patient and situation is unique, and deviations from protocol may be indicated. Like clinical problems, ethical problems require action for resolution. In most clinical settings, there is adequate time to identify and discuss the relevant ethical issues before decisions are made. However, this is often not the case in the (ED), where ethical problems may require immediate resolution, Arras, (1993).

To promote ethical decision making in these situations, a system to quickly analyze ethical concerns should be in place. McElroy, (2012) develop system can include consideration of the following issues:

- Who are the stakeholders?
- What is the chronology of events?
- What medical, social, and legal information is required to facilitate decision making?
- What is the best communication pathway to follow?
- What family values must be considered?
- Is there any consensus that exists with any of the person involved?

Public Guardians

The (ED) is unique from all other specialties in healthcare, and presents in a unique environment with distinct moral challenges. In order to respond appropriately to these ethical challenges, (ED) nurses are required to have knowledge of moral concepts and principles, and specialized moral reasoning skills. It becomes import then to identify and promote the moral attributes of those nurses in the (ED), McElroy, (2012).

Emergency nurses have a duty not only to their patients, but to the society in which they live. The nurse is responsible for informing the public, assisting in the allocation of resources in a just manner, opposing violence and promoting public health. It then becomes the responsibility of the ED nurse to participate in helping craft legislative, regulatory, institutional and educational pursuits that promote the safety of the patient and improve the quality of care, Arras, (1993).

Legal Aspects of Emergency Care

Emergency nurses interact with various aspects of the legal system throughout their career. Along with a relatively high likelihood of malpractice claims, emergency nurses often interact with police as they assess and treat trauma patients. They must also routinely deal with complicated legal issues such as advance directives, informed consent, protection of minors, mandatory reporting to health authorities, involuntary confinement, and compliance with federal laws, such as the Emergency Medical Treatment and Active Labor Act (EMTALA).

This part introduces the most common legal concepts that arise in emergency nursing It begins with a discussion of legal issues that develop during patient care, followed by a review of nurses interactions with the criminal justice system, providing care under the (EMTALA), and finally nursing malpractice.

The difficulty with any discussion of legal issues on a general level lies in the lack of consistency between the various legal systems. Broadly speaking, there are two main divisions' federal law and state law and each of these divisions is further subdivided into criminal, civil, and administrative sections. The laws governing any particular issue may differ depending on the particular state involved (or the federal government) and whether criminal, civil, or administrative rules apply. This inconsistency results in a certain ambiguity when discussing legal issues in a general sense, and mandates that nurses understand the specific laws that govern in their practice location.

Patient Care Legal Issues:

Informed consent

Certain legal issues begin the moment any patient enters the emergency department (ED); consent to treatment will likely be the first encountered. Competent patients have the right to refuse medical care. Conversely, nurses must obtain consent prior to procedures and treatment or risk a charge of battery. Informed consent is a process whereby the physician and patient discuss the risks, benefits, and alternatives to a given procedure or treatment. Although a patient's signature on an informed consent document may be prudent, it does not by itself meet the legal requirement of this process. To give informed consent, a patient must have sufficient information on which to base a decision about treatment.

The Legal determination of whether the nurse obtained informed consent varies from state to state, but relies on terms such as "appropriate" and "reasonable," thus leaving the issue far from clear in any given case.

Emergency exceptions to consent

Certain patients are unable to either consent to or refuse treatment based on their inability to participate meaningfully in the informed consent process. This typically occurs due to age, intoxication, underlying medical conditions, or acute changes in mental status. In documenting either informed consent or informed refusal, the nurse should make a notation of the patient's mental status and ability to consent or refuse. Nurses should refrain from referring to a patient's *competence*, as this is a legal term determined by a court, not a nurse. Rather, nurse s should refer to a patient's *decision-making capacity*.

When the patient cannot provide consent, all states have emergency exceptions to the consent process. Patients medically unable to express consent are presumed to consent to emergency stabilizing treatment. Intoxicated patients usually cannot provide consent, and fall under the emergency exception if the patient's decision making capacity is impaired.

Treatment of minors

Although most children cannot provide legal consent, all states have exemptions for children presenting to the (ED) without a parent or legal guardian. In these situations, the nurse can assume consent for any necessary stabilizing treatment. Additionally, federal law (EMTALA) requires that all children, whether accompanied by a consenting adult or not, receive a medical screening examination (MSE) and any necessary stabilizing treatment. In the rare case of a guardian or parent refusing to consent to life-saving treatment needed to stabilize a child, nurses should generally provide necessary treatment and obtain a court order after the fact.

Note that age by itself does not necessarily preclude the ability to consent, as there are conditions that allow minors to make their own medical decisions. These conditions vary from state to state, but often include the ability to consent without parental knowledge to treatment for sexually transmitted illnesses, pregnancy, drug and alcohol dependency, rape, and mental health concerns. Emancipated minors, individuals who the law recognizes as adults despite their age, also can consent or refuse treatment without parental involvement. Circumstances that allow minors to apply for legal emancipation vary, but often include active duty in the armed forces, marriage, pregnancy, and parenthood.

Legal aspects of emergency care informed refusal

Patients with decision-making capacity may withdraw consent for treatment at any time. This may occur prior to the involvement of a nurse, when patients leave the (ED) before being evaluated, or at some point thereafter when they either refuse treatment or leave against medical advice (AMA). When a patient withdraws consent, the nurse must again determine his or her capacity to do so. Patients with impaired decision making capacity cannot provide an informed refusal, and the nurse has a duty to provide stabilizing treatment.

The majority of patients who Leave (AMA) does have the capacity to make that decision, but every effort should be made to uncover the underlying reason for deciding to leave. Occasionally, patients don't understand the various treatment options or the potential for deterioration of their condition. While a signature on an (AMA) form may be helpful defending against litigation, a note in the nursing record detailing the patient's capacity and the process by which informed refusal occurred must supplement it.

Ideally, a family member should witness this discussion and sign the (AMA) form as well. Some patients will withdraw consent by simply leaving the (ED), known as "eloping." These patients constitute a dilemma for the nurse, as not all patients can "elope." For example, a 75-year-old patient with advanced Alzheimer's disease who wanders from the (ED) can neither "elope" nor provide informed refusal. Patients who "elope" should be discussed among involved care providers, including nursing, to determine if the patient reasonably appeared to have the capacity to withdraw consent and leave.

Advance directives

Often patients who do not have the capacity to consent have either a *do-not-resuscitate (DNR)* order or an *advance directive* that assists in determining their wishes. A (DNR) order usually relates only to the specific condition of cardiopulmonary arrest; that is, a patient does not wish intubation, chest compressions, or defibrillation should their heart stop.

Unfortunately, unless these orders are explicit, it often remains unclear as to what exactly the patient would want for conditions less severe than cardiac arrest. If the patient displays signs of impending respiratory failure, should they be intubated? If they are hypotensive from unstable ventricular tachycardia, should they be cardioverted? Generally, a (DNR) Order does not equal a "no care" or "comfort care only" order, and should be interpreted narrowly.

The advance directive and *durable power of attorney for health care* can be more explicit than a general (DNR), and may provide more direction to the emergency physician. One of the difficulties common in the (ED) arises when a patient is thought to have a (DNR) order by staff but no documentation exists. Unfortunately, the laws governing this situation vary from state to state, but written documentation or "actual knowledge" of (DNR) status must usually be present to withhold care. What constitutes "actual knowledge" may be debated in individual cases, but generally refers to situations in which a nurse has personally seen the advance directive even though it is not currently present.

Involuntary detainment

Another situation in which patients may lose the ability to consent involves psychiatric conditions, such as severe depression resulting in a suicide attempt. In California, for example, patients who are suicidal, homicidal, or gravely disabled may be detained up to 72 hours for psychiatric evaluation. During this period, these patients may receive treatment necessary to stabilize a condition brought on by a suicide attempt, and they may be restrained by physical or chemical means necessary to protect themselves or others. The procedure for involuntarily detaining a patient is complex, and nurse s should become thoroughly familiar with the laws in their area, carefully document the events that required detainment, and the methods of restraint implemented.

Privacy and confidentiality

A further common legal issue arising during patient care involves privacy and confidentiality issues. The Health Insurance Portability and Accountability Act (HIPAA) set forth standards for hospitals with regard to both personal privacy and the confidentiality of medical records. In general, medical information cannot be shared with a third party without the consent of the patient unless such information is necessary for nursing treatment. This includes releasing information to other family members, insurance companies, and employers. Exceptions to this exist, as all states have Mandatory reporting of victims of violence and of certain health conditions.

Emergency nurses must be careful to obtain consent from patients prior to speaking with other family members, friends, or employers. The requirements of (HIPAA) do not apply to information shared between health care providers for the sole purpose of medical evaluation and treatment. Patients unable to consent due to nursing conditions represent a special case that the courts have yet to fully explore.

Generally, nurses should proceed with the assumption of what a reasonable person in a similar situation would want with regard to confidentiality. For example, refusing to update a spouse on the status of a critically injured patient because the patient is unable to give consent for the release of medical information seems unreasonable. While the courts have yet to determine many issues in this area, they typically allow discretion when a nurse acts in the best interest of an incapacitated patient.

However, in cases of human immunodeficiency virus (HIV) infection, drug and alcohol intoxication, and mental illness, the risk of damage to the patient for a breach of confidentiality is often considered substantial, and the medical information of patients with these diagnoses should be shared cautiously, if at all. Similarly, certain medical information, such as drug or alcohol tests in the hands of civil authorities, may result in serious consequences to some patients. Nurses in most states may not provide the police medical information on a patient without that patient's consent or a court order (excluding mandatory reporting requirements).

Description of chest injury:

A chest injury is any form of physical injury to the chest including the ribs, heart and lungs (figure2). Typically chest injuries are caused by blunt mechanisms such as motor vehicle collisions or penetrating mechanisms such as stabbings. A chest injury can occur as the result of an accidental or deliberate penetration of a foreign object into the chest. This type of injury can also result from a blunt trauma, leading to chest wall injury (causing rib bruises, fracture, and lung or heart contusions). As mentioned before, Chest injuries account for (25%) of all deaths from traumatic injury. Approximately (60%) of all multisystem trauma victims have some types of chest or thoracic trauma (Owens, Chaudry, Eggersted & Smith, 2000).

Figure (1): Anatomical structure of the thoracic cavity.

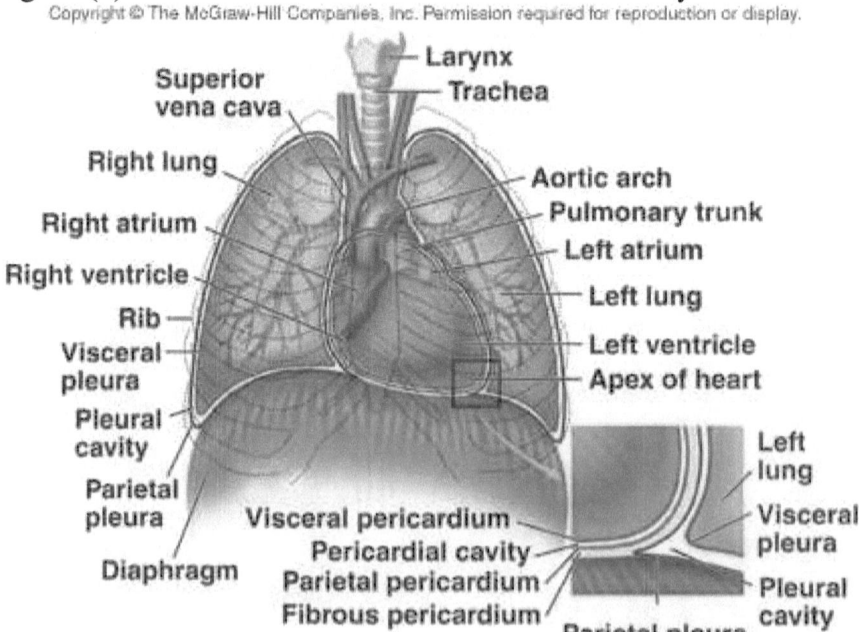

Pathophysiology of chest trauma

In The chest, anatomy and physiology are inextricably linked. The Normal functions of the thoracic organs depend on normal anatomic compartmentation. When these compartments are either opened to atmospheric pressure or filled with fluids, normal physiologic mechanisms are interrupted (figure3). An Injury to one thoracic organ alters the function of the other, Calhoon & Trinkle (1997). Injuries to the chest are often life-threatening and result in one or more of the following pathologic mechanisms:

1- **Hypoxemia** from disruption of the airway; injury to the lung parenchyma, rib cage, and respiratory musculature; massive hemorrhage; collapsed lung; and pneumothorax.

2- **Hypovolemia** from massive fluid loss from the great vessels, cardiac rupture, or hemothorax.

3- Cardiac failure from cardiac tamponade, cardiac contusion, or increased intrathoracic pressure these mechanisms frequently result in **ventilation and perfusion mismatching** leading to Acute Respiratory Failure (ARF), hypovolemic shock, and death, Kramer (2002).

Figure(3): Pathophysiology of chest trauma.

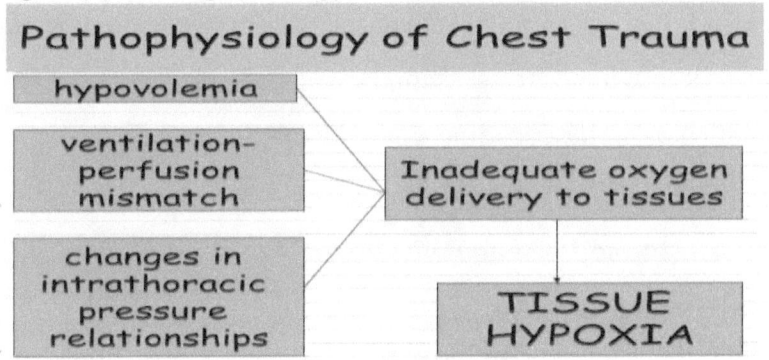

Assessment and Diagnostic Findings:

Time is critical in treating chest trauma. Therefore, it is essential to assess the patient immediately to determine the following:

• When the injury occurred

• Mechanism of injury

• Level of responsiveness

• Specific injuries

• Estimated blood loss

• Recent drug or alcohol use

• Pre-hospital treatment

The initial assessment of thoracic injuries includes assessment of the patient for airway obstruction, tension pneumothorax, open pneumothorax, massive hemothorax, flail chest, and cardiac tamponade. These injuries are life-threatening and need immediate treatment. Secondary assessment would include simple pneumothorax, hemothorax, pulmonary contusion, traumatic aortic rupture, tracheobronchial disruption, esophageal perforation, traumatic diaphragmatic injury, and penetrating wounds to the mediastinum (Owens, Chaudry, Eggerstedt & Smith, 2000). Although listed as secondary, these injuries may be life-threatening as well depending upon the circumstances.

The physical examination includes inspection of the airway, thorax, neck veins, and breathing difficulty. Specific include assessing the rate and depth of breathing for abnormalities, such as stridor, cyanosis, nasal flaring, use of accessory muscles, drooling, and overt trauma to the face, mouth, or neck. The chest should be auscultation assessed for symmetric movement, symmetry of breath sounds, open chest wounds, entrance or exit wounds, impaled objects, tracheal shift, distended neck veins, subcutaneous emphysema, and paradoxical chest wall motion. In addition, the chest wall should be assessed for bruising, petechiae, lacerations, and burns. The vital signs and skin color are assessed for signs of shock. The thorax is palpated for tenderness and crepitus (figure 4).

The initial diagnostic workup includes a chest x-ray (figure 5), CT scan, complete blood count, clotting studies, type and cross-match, electrolytes, oxygen saturation, arterial blood gas analysis, and ECG. The patient is completely undressed to avoid missing additional injuries that can complicate care. Many patients with injuries involving the chest have associated head and abdominal injuries that require attention. Ongoing assessment is essential to monitor the patient's response to treatment and to detect early signs of clinical deterioration.

Figure (4) palpation for tenderness   Figure (5) chest x-ray

Goals of Management:

The goals of treatment are to evaluate the patient's condition and to initiate aggressive resuscitation. An airway is immediately established with oxygen support and, in some cases, intubation and ventilator support. Re-establishing fluid volume and negative intra-pleural pressure and draining intra-pleural fluid and blood are essential.

The potential for massive blood loss and exsanguinations with blunt or penetrating chest injuries is high because of injury to the great blood vessels. Many patients die at the scene or are in shock by the time help arrives. Agitation and irrational and combative behavior are signs of decreased oxygen delivery to the cerebral cortex.

Strategies to restore and maintain cardiopulmonary function include ensuring an adequate airway and ventilation, stabilizing and re-establishing chest wall integrity, occluding any opening into the chest (open pneumothorax), and draining or removing any air or fluid from the thorax to relieve pneumothorax, hemothorax, or cardiac tamponade. Hypovolemia and low cardiac output must be corrected. Many of these treatment efforts, along with the control of hemorrhage, are usually carried out simultaneously at the scene of the injury or in the emergency department. Depending on the success of efforts to control the Hemorrhage in the emergency department, the patient may be taken immediately to the operating room. Principles of management are essentially those pertaining to care of the postoperative thoracic patient (figure 6).

Figure (6): stages of chest trauma initial management:

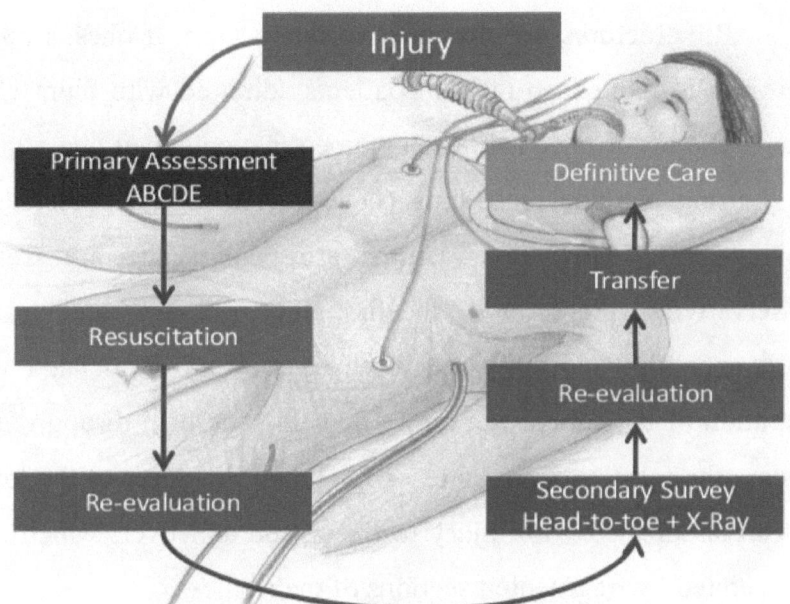

Common cardio-thoracic types of injury:

- Sternal and rib Fractures.
- Flail chest.
- Pulmonary contusion.
- Penetrating wound.
- Pneumothorax (Simple or Traumatic "tension, open").
- Hemothorax.
- Cardiac tamponade.
- Esophageal perforation.
- Subcutaneous emphysema (SCE).

Sternal and Rib Fractures

Rib fractures are the most common type of chest trauma, occurring in more than 60% of patients admitted with blunt chest injury. The occurrence is high in women, patients over age 50, and those using shoulder restraints (Owens, Chaudry, Eggerstedt & Smith, 2000). Most rib fractures are benign and are treated conservatively. Fractures of the first three ribs are rare but can result in a high mortality rate because they are associated with laceration of the subclavian artery or vein. The fifth through ninth ribs are the most common sites of fractures. Fractures of the lower ribs are associated with injury to the spleen and liver, which may be lacerated by fragmented sections of the rib.

Figure (7): Sternal and Rib Fractures:

Clinical Manifestations

The patient with sternal fractures has anterior chest pain, overlying tenderness, ecchymosis, crepitus, swelling, and the potential of a chest wall deformity. For the patient with rib fractures, clinical manifestations are similar: severe pain, point tenderness, and muscle spasm over the area of the fracture, which is aggravated by coughing, deep breathing, and movement. The area around the fracture may be bruised. To reduce the pain, the patient splints the chest by breathing in a shallow manner and avoids sighs deep breaths, coughing, and movement. This reluctance to move or breathe deeply results in diminished ventilation, collapse of unaerated alveoli (atelectasis), pneumonitis, and hypoxemia. Respiratory insufficiency and failure can be the outcomes of such a cycle.

Assessment and Diagnostic Findings

The patient with a sternal fracture must be closely evaluated for underlying cardiac injuries. A crackling, grating sound in the thorax (subcutaneous crepitus) may be detected with auscultation. The diagnostic workup may include a chest x-ray, rib films of a specific area, ECG, continuous pulse oximetry, and arterial blood gas analysis.

Management

Management of the patient with a sternal fracture is directed toward controlling pain, avoiding excessive activity, and treating any associated injuries. Surgical fixation is rarely necessary unless fragments are grossly displaced and pose a potential for further injury.

The goals of treatment for rib fractures are to control pain and to detect and treat the injury. fixation of multiple simple fractured ribs (3-12th.) by Elastic bandage or Plaster strapping only from midline (anterior) to midline (posterior) avoiding to encircle the whole circumference of the thoracic cage (figure 8). Sedation is used to relieve pain and to allow deep breathing and coughing. Care must be taken to avoid over sedation and suppression of the respiratory drive. Alternative strategies to relieve pain include an intercostal nerve block and ice over the fracture site; a chest binder may decrease pain on movement. Usually the pain abates in 5 to 7 days, and discomfort can be controlled with epidural analgesia, patient-controlled analgesia, or nonopioid analgesia. Most rib fractures heal in 3 to 6 weeks. The patient is monitored closely for signs and symptoms of associated injuries.

Figure (8): rip fracture external fixation:

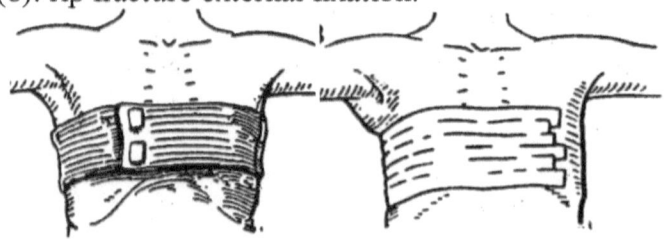

Flail Chest:

Flail chest is frequently a complication of blunt chest trauma from a steering wheel injury. It usually occurs when three or more adjacent ribs (multiple contiguous ribs) are fractured at two or more sites, resulting in free-floating rib segments. It may also result as a combination fracture of ribs and costal cartilages or sternum. As a result, the chest wall loses stability and there is subsequent respiratory impairment and usually severe respiratory distress (figure 9).

Figure (9) Flail chest (paradoxical movement)

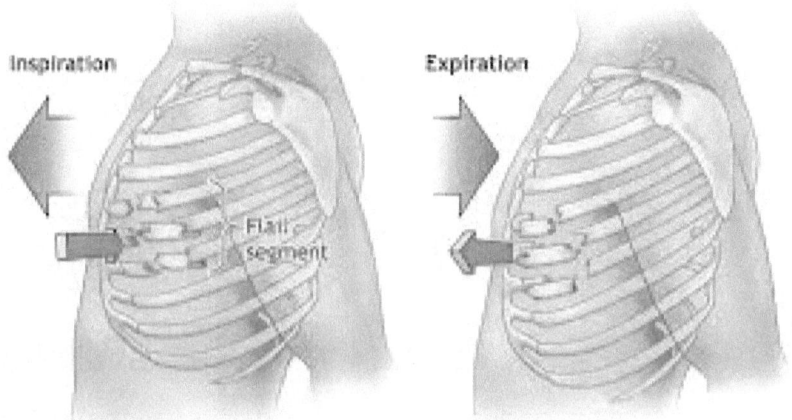

Pathophysiology:

During inspiration, as the chest expands, the detached part of the rib segment (flail segment) moves in a paradoxical manner (pendelluft movement) in that it is pulled inward during inspiration, reducing the amount of air that can be drawn into the lungs. On expiration, because the intrathoracic pressure exceeds atmospheric pressure, the flail segment bulges outward, impairing the patient's ability to exhale. The mediastinum then shifts back to the affected side (Figure 8).

This paradoxical action results in increased dead space, a reduction in alveolar ventilation, and decreased compliance. Retained airway secretions and atelectasis frequently accompany flail chest. The patient has hypoxemia, and if gas exchange is greatly compromised, respiratory acidosis develops as a result of (CO_2) retention. Hypotension, inadequate tissue perfusion, and metabolic acidosis often follow as the paradoxical motion of the mediastinum decreases cardiac output.

Management:

As with rib fracture, treatment of flail chest is usually supportive. Management includes providing ventilatory support, clearing secretions from the lungs, and controlling pain. The specific management depends on the degree of respiratory dysfunction. If only a small segment of the chest is involved, the objectives are to clear the airway through positioning, coughing, deep breathing, and suctioning to aid in the expansion of the lung, and to relieve pain by intercostal nerve blocks (figure 10), high thoracic epidural blocks, or cautious use of intravenous opioids. For mild to moderate flail chest injuries, the underlying pulmonary contusion is treated by monitoring fluid intake and appropriate fluid replacement, while at the same time relieving chest pain. Pulmonary physiotherapy focusing on lung volume expansion and secretion management techniques are performed. The patient is closely monitored for further respiratory compromise.

Figure (10): intercostal nerve blocks.

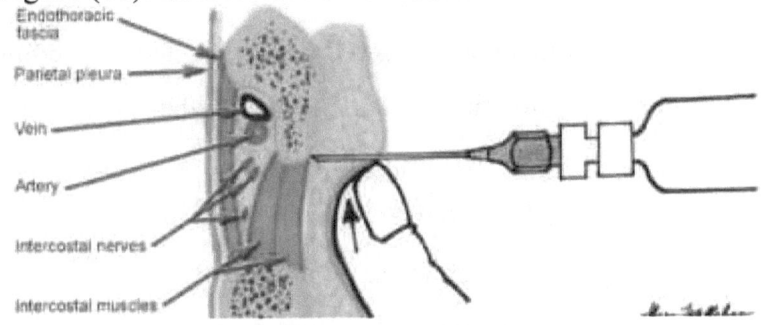

When a severe flail chest injury is encountered, (ETT) intubation and mechanical ventilation are required to provide internal pneumatic stabilization of the flail chest and to correct abnormalities in gas exchange. This helps to treat the underlying pulmonary contusion, serves to stabilize the thoracic cage to allow the fractures to heal, and improves alveolar ventilation and intrathoracic volume by decreasing the work of breathing.

Differing modes of ventilation are used depending on the patient's underlying disease and specific needs. In rare circumstances, surgery may be required to more quickly stabilize the flail segment. This may be used in the patient who is difficult to ventilate or the high-risk patient with underlying lung disease who may be difficult to wean from mechanical ventilation. Regardless of the type of treatment, the patient is carefully monitored by serial chest x-rays, arterial blood gas analysis, pulse oximetry, and bedside pulmonary function monitoring. Pain management is key to successful treatment. Patient-controlled analgesia, intercostal nerve blocks, epidural analgesia, and intrapleural administration of opioids may be used to control thoracic pain.

45

Pulmonary Contusion:

Pulmonary contusion is observed in about 20% of adult patients with multiple traumatic injuries and in a higher percentage of children due to increased compliance of the chest wall. It is defined as damage to the lung tissues resulting in hemorrhage and localized edema (figure 11). It is associated with chest trauma when there is rapid compression and decompression to the chest wall (ie, blunt trauma). It may not be evident initially on examination but will develop in the posttraumatic period.

Figure (11): Bilateral Pulmonary Contusion.

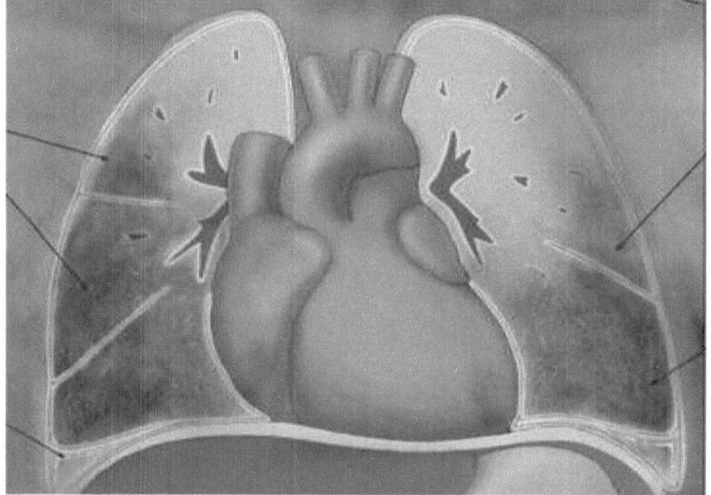

Pathophysiology

The primary pathologic defect is an abnormal accumulation of fluid in the interstitial and intra-alveolar spaces. It is thought that injury to the lung parenchyma and its capillary network results in a leakage of serum protein and plasma. The leaking serum protein exerts an osmotic pressure that enhances loss of fluid from the capillaries. Blood, edema, and cellular debris (from cellular response to injury) enter the lung and accumulate in the bronchioles and alveolar surface, where they interfere with gas exchange. An increase in pulmonary vascular resistance and pulmonary artery pressure occurs. The patient has hypoxemia and carbon dioxide retention. Occasionally, a contused lung occurs on the other side of the point of body impact; this is called a contrecoup contusion.

Clinical Manifestations:

Pulmonary contusion may be mild, moderate, or severe. The clinical manifestations vary from tachypnea, tachycardia, pleuritic chest pain, hypoxemia, and blood-tinged secretions to more severe tachypnea, tachycardia, crackles, frank bleeding, severe hypoxemia, and respiratory acidosis. Changes in sensorium, including increased agitation or combative irrational behavior, may be signs of hypoxemia. In addition, the patient with moderate pulmonary contusion has a large amount of mucus, serum, and frank blood in the tracheobronchial tree; the patient often has a constant cough but cannot clear the secretions. A patient with severe pulmonary contusion has the signs and symptoms of (ARDS); these may include central cyanosis, agitation, combativeness, and productive cough with frothy, bloody secretions.

Assessment and Diagnostic Findings:

The efficiency of gas exchange is determined by pulse oximetry and arterial blood gas measurements. Pulse oximetry is also used to measure oxygen saturation continuously. The chest x-ray may show pulmonary infiltration. The initial chest x-ray may show no changes; in fact, changes may not appear for 1 or 2 days after the injury.

Management:

Treatment priorities include maintaining the airway, providing adequate oxygenation, and controlling pain. In mild pulmonary contusion, adequate hydration via intravenous fluids and oral intake is important to mobilize secretions. However, fluid intake must be closely monitored to avoid hypervolemia. Volume expansion techniques, postural drainage, physiotherapy including coughing and endotracheal suctioning are used to remove the secretions. Pain is managed by intercostal nerve blocks or by opioids via patient-controlled analgesia or other methods.

Usually, antimicrobial therapy is administered because the damaged lung is susceptible to infection. Supplemental oxygen is usually given by mask or cannula for (24 to 36 hours). The patient with moderate pulmonary contusion may require bronchoscopy to remove secretions; intubation and mechanical ventilation with (PEEP) may also be necessary to maintain the pressure and keep the lungs inflated. Diuretics may be given to reduce edema. A nasogastric tube is inserted to relieve gastrointestinal distention. The patient with severe contusion may develop respiratory failure and may require aggressive treatment with endotracheal intubation and ventilatory support (figure 12), diuretics, and fluid restriction. Colloids and crystalloid solutions may be used to treat hypovolemia. Antimicrobial medications may be prescribed for the treatment of pulmonary infection. This is a common complication of pulmonary contusion (especially pneumonia in the contused segment), because the fluid and blood that extravasates into the alveolar and interstitial spaces serve as an excellent culture medium.

Figure (12): Endotracheal Intubation and Ventilatory Support

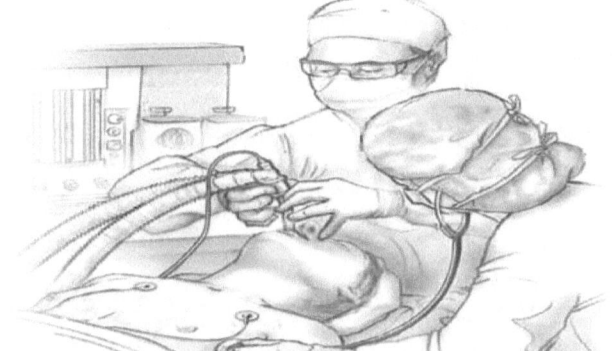

Penetrating Wound:

Gunshot and stab wounds are the most common types of penetrating chest trauma. They are classified according to their velocity. Stab wounds are generally considered of low velocity because the weapon destroys a small area around the wound. Knives and switchblades cause most stab wounds. The appearance of the external wound may be very deceptive, because pneumothorax, hemothorax, lung contusion, and cardiac tamponade, along with severe and continuing hemorrhage, can occur from any small wound, even one caused by a small-diameter instrument such as an ice pick (figure13).

Figure (13): Penetrating wound extension.

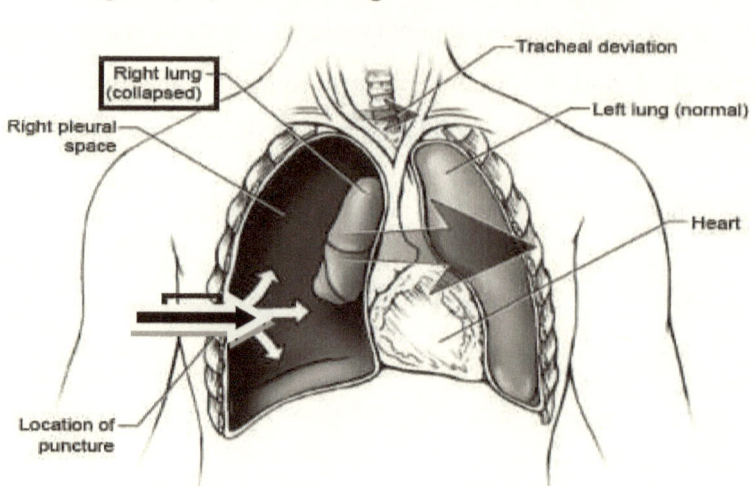

Gunshot wounds to the chest may be classified as of low, medium, or high velocity. The factors that determine the velocity and resulting extent of damage include the distance from which the gun was fired, the caliber of the gun, and construction and size of the bullet. A gunshot wound can produce a variety of pathophysiologic changes. A bullet can cause damage at the site of penetration and along its pathway. It also may ricochet off bony structures and damage the chest organs and great vessels. If the diaphragm is involved in either a gunshot wound or a stab wound, injury to the chest cavity must be considered.

Management:

The objective of immediate management is to restore and maintain cardiopulmonary function. After an adequate airway is ensured and ventilation is established, the patient is examined for shock and intrathoracic and intra-abdominal injuries.

The patient is undressed completely so that additional injuries will not be missed. There is a high risk for associated intra-abdominal injuries with stab wounds below the level of the fifth anterior intercostals space. Death can result from exsanguinating hemorrhage or intra-abdominal sepsis. After the status of the peripheral pulses is assessed, a large-bore intravenous line is inserted. The diagnostic workup includes a chest x-ray, chemistry profile, arterial blood gas analysis, pulse oximetry, and ECG. Blood typing and cross-matching are done in case blood transfusion is required. An indwelling catheter is inserted to monitor urinary output. A nasogastric tube is inserted to prevent aspiration, minimize leakage of abdominal contents, and decompress the gastrointestinal tract. Shock is treated simultaneously with colloid solutions, crystalloids, or blood, as indicated by the patient's condition.

Other diagnostic procedures are carried out as needed of the patient (eg, CT scans of chest, flat plate x-ray of the abdomen, abdominal tap to check for bleeding). A chest tube is inserted into the pleural space in most patients with penetrating wounds of the chest to achieve rapid and continuing re-expansion of the lungs. The insertion of the chest tube frequently results in a complete evacuation of the blood and air. The chest tube also allows early recognition of continuing intra-thoracic bleeding, which would make surgical exploration necessary. If the patient has a penetrating wound of the heart and great vessels, the esophagus, or the tracheobronchial tree, surgical intervention is required.

Pneumothorax

Pneumothorax occurs when the parietal or visceral pleura are breached and the pleural space is exposed to positive atmospheric pressure. Normally the pressure in the pleural space is negative or subatmospheric compared to atmospheric pressure; this negative pressure is required to maintain lung inflation. When either pleura is breached, air enters the pleural space, and the lung or a portion of it collapses. Types of pneumothorax include simple, traumatic, and tension pneumothorax.

Simple Pneumothorax

A simple, or spontaneous, pneumothorax occurs when air enters the pleural space through a breach of either the parietal or visceral pleura. Most commonly this occurs as air enters the pleural space through the rupture of a bleb or a bronchopleural fistula. A spontaneous pneumothorax may occur in an apparently healthy person in the absence of trauma due to rupture of an air-filled bleb, or blister, on the surface of the lung, allowing air from the airways to enter the pleural cavity. It may be associated with diffuse interstitial lung disease and severe emphysema (figure 14).

Figuer (14) spontaneous pneumothorax

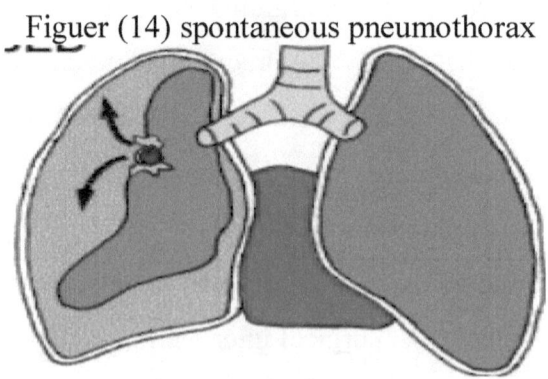

Traumatic Pneumothorax

Traumatic pneumothorax occurs when air escapes from a laceration in the lung itself and enters the pleural space or enters the pleural space through a wound in the chest wall. It can occur with blunt trauma (eg, rib fractures) or penetrating chest trauma. It may also occur from abdominal trauma (eg, stab wounds or gunshot wounds to the abdomen) and from diaphragmatic tears. Traumatic pneumothorax may occur with invasive thoracic procedures (ie, thoracentesis, transbronchial lung biopsy, insertion of a subclavian line) in which the pleura is inadvertently punctured, or with barotrauma from mechanical ventilation (figure 15).

Figure (15): Types of Traumatic pneumothorax

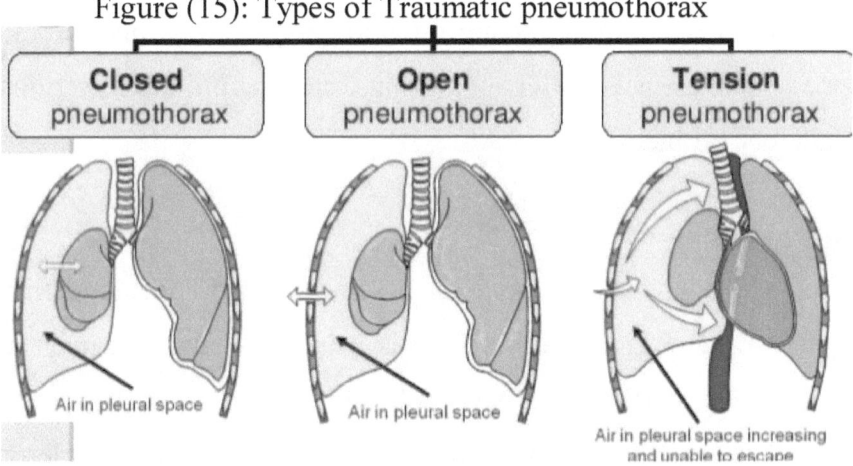

Traumatic pneumothorax resulting from major injury to the chest is often accompanied by hemothorax (collection of blood in the pleural space resulting from torn intercostal vessels, lacerations of the great vessels, and lacerations of the lungs). Often both blood and air are found in the chest cavity (hemo-pneumothorax) after major trauma. Chest surgery can cause what is classified as a traumatic pneumothorax as a result of the entry into the pleural space and the accumulation of air and fluid in the pleural space.

Open pneumothorax:

Open pneumothorax is one form of traumatic pneumothorax. It occurs when a wound in the chest wall is large enough to allow air to pass freely in and out of the thoracic cavity with each attempted respiration. Because the rush of air through the hole in the chest wall produces a sucking sound, such injuries are termed sucking chest wounds. In such patients, not only does the lung collapse, but the structures of the mediastinum (heart and great vessels) also shift toward the uninjured side with each inspiration and in the opposite direction with expiration. This is termed mediastinal flutter or swing, and it produces serious circulatory problems.

Clinical Manifestations:

The signs and symptoms associated with pneumothorax depend on its size and cause. Pain is usually sudden and may be pleuritic. The patient may have only minimal respiratory distress with slight chest discomfort and tachypnea with a small simple or uncomplicated pneumothorax. If the pneumothorax is large and the lung collapses totally, acute respiratory distress occurs. The patient is anxious, has dyspnea and air hunger, has increased use of the accessory muscles, and may develop central cyanosis from severe hypoxemia. Severe chest pain may occur, accompanied by tachypnea, decreased movement of the affected side of the thorax, a tympanic sound on percussion of the chest wall, and decreased or absent breath sounds and tactile fremitus on the affected side.

Medical Management

Medical management of pneumothorax depends on its cause and severity. The goal of treatment is to evacuate the air or blood from the pleural space. A small chest tube (28 French) is inserted near the second intercostal space; this space is used because it is the thinnest part of the chest wall, minimizes the danger of contacting the thoracic nerve, and leaves a less visible scar (figure 16).

Figure (16): pneumothorax management by chest tube

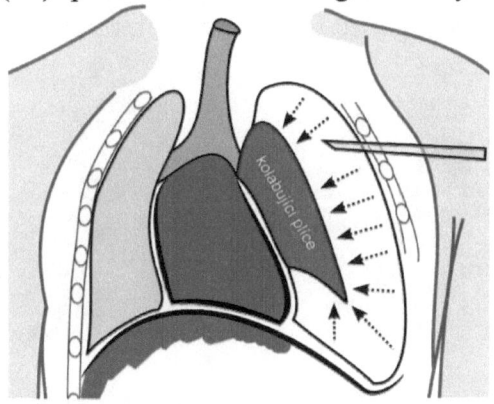

If the patient also has a hemothorax, a large-diameter chest tube (32 French or greater) is inserted, usually in the fourth or fifth intercostal space at the midaxillary line. The tube is directed posteriorly to drain the fluid and air. Once the chest tube or tubes are inserted and suction is applied (usually to 20 mm Hg suction), effective decompression of the pleural cavity (drainage of blood or air) occurs.

If an excessive amount of blood enters the chest tube in a relatively short period, an autotransfusion may be needed. This technique involves taking the patient's own blood that has been drained from the chest, filtering it, and then transfusing it back into the patient's vascular system. In such an emergency, anything may be used that is large enough to fill the chest wound a towel, a handkerchief, or the heel of the hand. If conscious, the patient is instructed to inhale and strain against a closed glottis. This action assists in reexpanding the lung and ejecting the air from the thorax. In the hospital, the opening is plugged by sealing it with gauze impregnated with petrolatum (figure 17).

Figure (17): Sealed bandage gauze for open pneumothorax

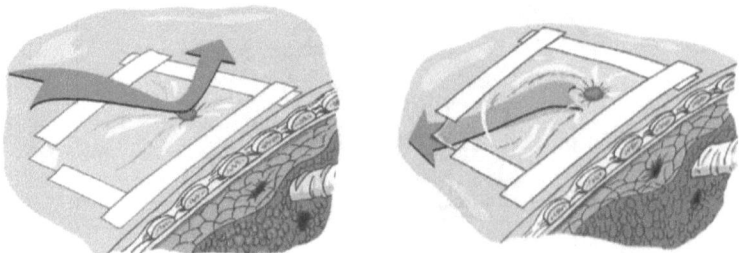

Usually, a chest tube connected to water-seal drainage is inserted to permit air and fluid to drain. Antibiotics usually are prescribed to combat infection from contamination. The severity of open pneumothorax depends on the amount and rate of thoracic bleeding and the amount of air in the pleural space. The pleural cavity can be decompressed by needle aspiration (thoracentesis) or chest tube drainage of the blood or air. The Lung is then able to re-expand and resume the function of gas exchange. As a rule of thumb, the chest wall is opened surgically (thoracotomy) when more than 1,500 mL of blood is aspirated initially by thoracentesis (or is the initial chest tube output) or when chest tube output continues at greater than 200 mL/hour. The urgency with which the blood must be removed is determined by the respiratory compromise. An emergency thoracotomy may also be performed in the emergency department if there is suggested cardiovascular injury secondary to chest or penetrating trauma.

Closed Pneumothorax

Closed pneumothorax may occur if a pneumothorax occurs suddenly or for no known reason, it is called a closed or spontaneous pneumothorax this condition most often strikes tall, thin men between the ages of 20 to 40. In addition, people with lung disorders, such as emphysema, cystic fibrosis, and tuberculosis, are at higher risk for closed pneumothorax. Traumatic closed pneumothorax is the result of accident or injury due to medical procedures performed to the chest cavity, such as thoracentesis or mechanical ventilation. In this type of pneumothorax, air enters the chest cavity, but cannot escape. This greatly increased pressure in the pleural space causes the lung to collapse completely, compresses the heart, and pushes the heart and associated blood vessels toward the unaffected side.

Clinical Manifestations:

The symptoms of pneumothrax depend on how much air enters the chest, how much the lung collapses, and the extent of lung disease. Spontaneous pneumothorax is characterized by dull, sharp, or stabbing chest pain that begins suddenly and becomes worse with deep breathing or coughing. Other symptoms are shortness of breath, rapid breathing, abnormal breathing movement (that is, little chest wall movement when breathing), and cough. There is marked anxiety, distended neck veins, weak pulse, decreased breath sounds on the affected side, and a shift of the mediastinum to the opposite side.

Diagnosis:

To diagnose closed pneumothorax, it is necessary for the health care provider to listen to the chest (auscultation) during a physical examination by using a stethoscope, the physician may note that one part of the chest does not transmit the normal sounds of breathing. A chest x ray will show the air pocket and the collapsed lung. An electrocardiogram (ECG) will be performed to record the electrical impulses that control the heart's activity. Blood samples may be taken to check for the level of arterial blood gases.

Treatment

A small pneumothorax may resolve on its own, but most require medical treatment. The object of treatment is to remove air from the chest and allow the lung to re-expand. This is done by inserting a needle and syringe (if the pneumothorax is small) or chest tube through the chest wall. This allows the air to escape without allowing any air back in. The lung will then re-expand itself within a few days. Surgery may be needed for repeat occurrences.

Tension Pneumothorax

A tension pneumothorax occurs when air is drawn into the pleural space from a lacerated lung or through a small hole in the chest wall. It may be a complication of other types of pneumothorax. In contrast to open pneumothorax, the air that enters the Chest cavity with each inspiration is trapped; it cannot be expelled during expiration through the air passages or the hole in the chest wall. In effect, a one-way valve or ball valve mechanism occurs where air enters the pleural space but cannot escape. With each breath, tension (positive pressure) is increased within the affected pleural space. This causes the lung to collapse and the heart, the great vessels, and the trachea to shift toward the unaffected side of the chest (mediastinal shift). Both respiration and circulatory function are compromised because of the increased intrathoracic pressure. The increased intrathoracic pressure decreases venous Return to the heart, causing decreased cardiac output and impairment of peripheral circulation. In extreme cases, the pulse may be undetectable this is known as pulseless electrical activity.

Clinical Manifestations

The clinical picture is one of air hunger, agitation, increasing hypoxemia, central cyanosis, hypotension, tachycardia, and profuse diaphoresis.

Medical Management:

If a tension pneumothorax is suspected, the patient should immediately be given a high concentration of supplemental oxygen to treat the hypoxemia, and pulse oximetry should be used to monitor oxygen saturation. In an emergency situation, a dressing called the "Asherman seal" should be utilized, as it appears to be more effective than a standard "three-sided" dressing. The Asherman seal is a specially designed device that adheres to the chest wall and, through a valve-like mechanism, allows air to escape but not to enter the chest (figure 18).

Figure (18): Asherman seal dressing.

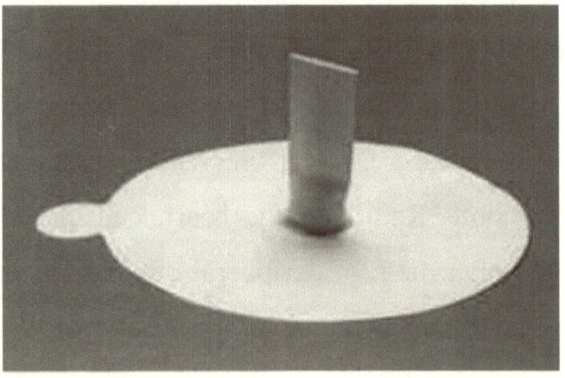

Tension pneumothorax can be decompressed or quickly converted to a simple pneumothorax by inserting a large-bore needle (14-gauge) at the second intercostals space, midclavicular line on the affected side. This relieves the pressure and vents the positive pressure to the external environment. Chest tube is then inserted and connected to suction to remove the remaining air and fluid, re-establish the negative pressure, and re-expand the lung. If the lung re-expands and air leakage from the lung parenchyma stops, further drainage may be unnecessary. If a prolonged air leak continues despite chest tube drainage to underwater seal, surgery may be necessary to close the leak.

Hemothorax

Hemothorax is the presence of blood in the pleural space (figure 19). The source of blood may be the chest wall, lung parenchyma, heart, or great vessels. Although some authors' state that a hematocrit value of at least (50%) is necessary to differentiate a hemothorax from a bloody pleural effusion, most do not agree on any specific distinction. Hemothorax is usually a consequence of blunt or penetrating trauma. Much less commonly, it may be a complication of disease, may be iatrogenically induced,or may develop spontaneously.

Fgure (19): Hemothorax

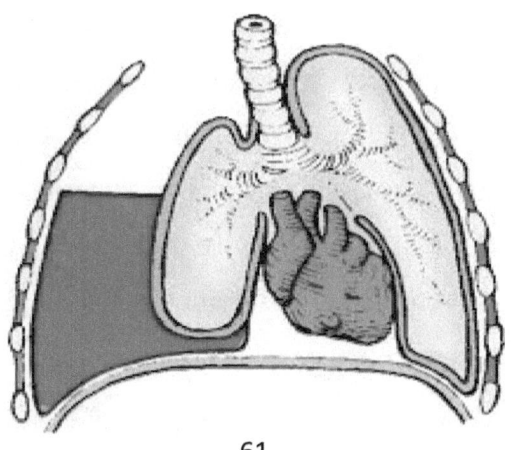

Pathophysiology

Bleeding into the pleural space can occur with virtually any disruption of the tissues of the chest wall and pleura or the intrathoracic structures. The physiologic response to the development of a hemothorax is manifested in two major areas: hemodynamic and respiratory. Hemodynamic changes vary, depending on the amount of bleeding and the rapidity of blood loss. Blood loss of up to (750 mL) in a (70-kg) man should cause no significant hemodynamic change. Loss of (750-1500) mL in the same individual will cause the early symptoms of shock (ie, tachycardia, tachypnea, and a decrease in pulse pressure). Significant signs of shock with signs of poor perfusion occur with loss of blood volume of 30% or more (1500-2000 mL). Because the pleural cavity of a 70-kg man can hold 4 L of blood or more, exsanguinating hemorrhage can occur without external evidence of blood loss.

In relation to the Respiratory response, the space-occupying effect of a large accumulation of blood within the pleural space may hamper normal respiratory movement. In trauma cases, abnormalities of ventilation and oxygenation may result, especially if associated with injuries to the chest wall. A large enough collection of blood causes the patient to experience dyspnea and may produce the clinical finding of tachypnea. The volume of blood required to produce these symptoms in a given individual varies depending on a number of factors, including organs injured, severity of injury, and underlying pulmonary and cardiac reserve.

Clinical Manifestations

Hemothorax tends to occur following blunt or penetrating trauma to the thorax or thoracoabdominal area. It may also follow thoracic surgery, or may be spontaneous. Chest pain, dyspnea, and tachypnea are common presenting features. Other symptoms of hemothorax are dependent on the mechanism of injury, but may include:

- Cyanosis
- Decreased or absent breath sounds on affected side
- Tracheal deviation to unaffected side
- Dull resonance on percussion
- Unequal chest rise
- Tachycardia
- Hypotension
- Pale, cool, clammy skin
- Possible subcutaneous emphysema
- Narrowing pulse pressure

Diagnosis

Chest radiography is the preferred means of initial diagnosis for hemothorax. Upright radiography is preferred but supine films may be taken when upright radiography is not feasible due to the clinical situation. Tube thoracostomy may be done prior to imaging when patients have sustained blunt or penetrating thoracic trauma and display unstable hemodynamics, have respiratory failure with absent or decreased breath sounds, show tracheal deviation, or have serious penetrating injuries. In upright radiography, hemothorax is suggested by blunting of the costophrenic angle or partial or complete opacification of the hemithorax, in which the lateral side of the chest appears bright and the lung appears pushed away toward the center; the air-filled lung normally appears as a dark space on radiographic film. In the case of a small hemothorax, several hundred milliliters of blood can be hidden by the diaphragm and abdominal viscera. In supine patients, signs of hemothorax may also be subtle on radiographic film, because the blood will layer in the pleural space, and can be seen as a haziness in one half of the thorax relative to the other side.

Ultrasonography is also used for detection of hemothorax and other pleural effusions, particularly in the critical care and trauma settings, because it provides rapid, reliable results in order to make a diagnosis in an emergency situation.Computed tomography (CT or CAT) scans can detect much smaller amounts of fluid than chest radiography, but computed tomography is not a primary method of diagnosis within the trauma setting, due to the time required for imaging, the requirement that a patient remain supine, and the need to transport a critically ill patient to the scanner.

Management:

A hemothorax is managed by removing the source of bleeding and by draining the blood already in the thoracic cavity. Blood in the cavity can be removed by inserting a drain (chest tube) in a procedure called a tube thoracostomy. Generally, the thoracostomy tube is placed between the ribs in the sixth or seventh intercostal space at the mid-axillary line. Usually the lung will expand and the bleeding will stop after a chest tube is inserted. The blood in the chest can thicken as the clotting cascade is activated when the blood leaves the blood vessels and comes into contact with the pleural surface, injured lung or chest wall, or with the chest tube. As the blood thickens, it can clot in the pleural space (leading to a retained hemothorax) or within the chest tube, leading to chest tube clogging or occlusion. Chest tube clogging or occlusion can lead to worse outcomes as it prevents adequate drainage of the pleural space, contributing to the problem of retained hemothorax. In this case, patients can be hypoxic, short of breath, , the retained hemothorax can become infected (empyema).

Retained hemothorax occurs when blood remains in the pleural space, and is a risk factor for the development of complications, including the accumulation of pus in the pleural space and fibrothorax. It is treated by inserting a second chest tube or by drainage by video-assisted thoracoscopy (figure 20). Fibrolytic therapy has also been studied as a treatment.

Figure (20): video-assisted thoracoscopy

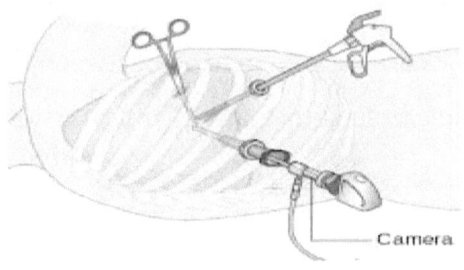

Camera

When hemothorax is treated with a chest tube, it is important that it maintain its function so that the blood cannot clot in the chest or the tube. If clogging occurs, internal chest tube clearing can be performed using an open or closed technique. Manual manipulation, which may also be called milking, stripping, or tapping, of chest tubes is commonly performed to maintain an open tube, but no conclusive evidence has demonstrated that any of these techniques are more effective than the others, or that they improve chest tube drainage. In some cases bleeding continues and surgery is necessary to stop the source of bleeding. For example, if the hemothorax was caused by aortic rupture in high energy trauma, surgical intervention is mandatory.

Physiologic resolution of hemothorax

Blood that enters the pleural cavity is exposed to the motion of the diaphragm, lungs, and other intrathoracic structures. This results in some degree of defibrination of the blood so that incomplete clotting occurs. Within several hours of cessation of bleeding, lysis of existing clots by pleural enzymes begins. Lysis of red blood cells results in a marked increase in the protein concentration of the pleural fluid and an increase in the osmotic pressure within the pleural cavity. This elevated intrapleural osmotic pressure produces an osmotic gradient between the pleural space and the surrounding tissues that favors transudation of fluid into the pleural space.

Cardiac Tamponade

Cardiac tamponade is an accumulation of massive bleeding in the pericardial sac compressing cardiac venous and arterial sides causing impairment of cardiac filling and evacuation leading to venous congestion and systemic hypotension (figure 21). The Pericardial sac is formed of 3 layers; Visceral layer a monolayer covering cardiac surface (Epicardium), Parietal layer bi-layered parietal (outer fibrous and inner smooth).

Figure (21): Cardiac tamponade (accumulation blood in pericardial sac)

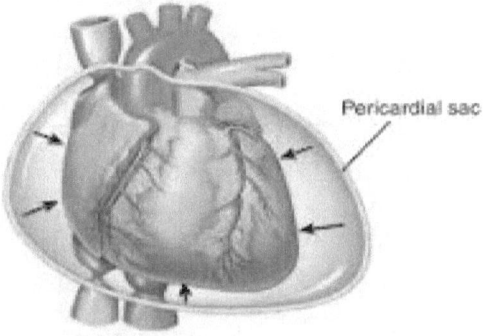

Causes:

Cardiac tamponade is caused by a large or uncontrolled pericardial effusion, i.e. the buildup of fluid inside the pericardium. This commonly occurs as a result of chest trauma (both blunt and penetrating), but can also be caused by myocardial rupture, cancer, uremia, pericarditis, or cardiac surgery, and rarely occurs during retrograde aortic dissection, or whilst the patient is taking anticoagulant therapy. The effusion can occur rapidly (as in the case of trauma or myocardial rupture), or over a more gradual period of time (as in cancer). The fluid involved is often blood, but pus is also found in some circumstances.

67

Causes of increased pericardial effusion include hypothyroidism, physical trauma (either penetrating trauma involving the pericardium or blunt chest trauma), pericarditis (inflammation of the pericardium), iatrogenic trauma (during an invasive procedure), and myocardial rupture. One of the most common causes is heart surgery, when post operative bleeding fails to be cleared by clogged chest tubes.

Pathophysiology:

Massive accumulation of pericardial sac fluids leads to compression on venous structures (SVC, IVC, atria) that cause venous return & ventricular filling, arterial structures (Aorta) which decrease Stroke Volume & Cardiac Output, and Coronary Vessels Myocardial performance & Cardiac Output. *Rate* of Accumulation is more important than *Amount accumulated*. Pericardial sac can accommodate big amounts of fluids without serious sequelae if accumulated gradually over a long time (eg: Uraemic patients). However, rapid (acute) accumulation of fluids (as little as 150 mls) can be fatal.

Clinical manifestation

Anxiety and restlessness, low blood pressure, weakness, chest pain radiating to your neck, shoulders, or back, trouble breathing or taking deep breaths, rapid breathing, discomfort that's relieved by sitting or leaning for war fainting, dizziness, and loss of consciousness.

Diagnosis

Serum Hemoglobin Continuously reduced (despite correction by blood transfusion) with or without Cyanosis. (ECG) Low-voltage, Frequent Extra-systoles or Ventricular Arrhythmias (mini-waves) despite controlled blood chemistry/electrolytes. Plain X-ray (PA view) Enlarged cardiac silhouette, flask-shaped heart. Echocardiography claimed to be sensitive and diagnostic.

Management:

Initial treatment given will usually be supportive in nature, for example administration of oxygen, and monitoring. There is little care that can be provided pre-hospital other than general treatment for shock. Prompt diagnosis and treatment is the key to survival with tamponade. Some pre-hospital providers will have facilities to provide pericardiocentesis (figure 22), which can be life-saving. If the patient has already suffered a cardiac arrest, pericardiocentesis alone cannot ensure survival, and so rapid evacuation to a hospital is usually the more appropriate course of action.

Figure (22): pericardiocentesis

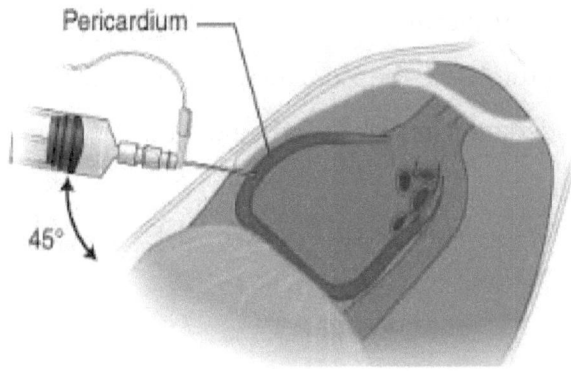

Pericardiocentesis procedure is insertion of a needle through the skin and into the pericardium and aspirating fluid under ultrasound guidance preferably. This can be done laterally through the intercostal spaces, usually the fifth, or as a subxiphoid approach (figure 22). A left parasternal approach begins 3 to 5 cm left of the sternum to avoid the left internal mammary artery, in the 5th intercostal space. Often, a cannula is left in place during resuscitation following initial drainage so that the procedure can be performed again if the need arises. If facilities are available, an emergency pericardial window may be performed instead, during which the pericardium is cut open to allow fluid to drain. Following stabilization of the patient, surgery is provided to seal the source of the bleed and mend the pericardium.

In heart surgery patients post op, the nurses monitor the amount of chest tube drainage. If the drainage volume drops off, and the blood pressure goes down, this can suggest tamponade due to chest tube clogging. In that case, the patient is taken back to the operating room for an emergency reoperation. If aggressive treatment is offered immediately and no complications arise (shock, AMI or arrhythmia, heart failure, aneurysm, carditis, embolism, or rupture), or they are dealt with quickly and fully contained, then adequate survival is still a distinct possibility.

Esophageal perforation

An esophageal perforation is a hole in the esophagus. The esophagus is the tube food passes through as it goes from the mouth to the stomach. The contents of the esophagus can pass into the area surrounding area in the chest (mediastinum), when there is a hole in the esophagus (figure 23). This often results in infection of the mediastinum (mediastinitis). The most common cause of an esophageal perforation is injury during a medical procedure. However, the use of flexible instruments has made this problem rare. The esophagus may also become perforated as the result of: A tumor, gastric reflux with ulceration, previous surgery on the esophagus, and swallowing a foreign object or caustic chemicals, such as household cleaners, disk batteries, and battery acid.

Figure (23): Esophageal perforation.

71

Clinical manifestation:

The main symptom is pain when the problem first occurs. A perforation in the middle or lower most part of the esophagus may cause: swallowing problems, chest pain, tachypnea, fever, hypotension, rapid heart rate, neck pain or stiffness.

You may have a chest x-ray to look for:

Diagnostic Procedures:

An imaging test, such as an X-ray, CT scan, or upper endoscopy to check for signs of esophageal perforation. These tests are used to look in the chest for air bubbles and abscesses, which are sacs filled with pus. They can also help your doctor see whether fluid has leaked out of your esophagus and into your lungs.

Management:

Treatment of the perforation must be done as quickly as possible to prevent infection. The earlier you get treatment, the better. Ideally, you should receive treatment within 24 hours of diagnosis. The fluid that leaks out of the hole in your esophagus can become trapped in the tissue between your lungs. This area is called the mediastinum. It's located behind your breastbone. The accumulation of fluid there can cause breathing difficulties and lung infections. A permanent stricture, or narrowing of the esophagus, can develop if your esophageal perforation isn't treated right away. This condition can make swallowing and breathing more difficult.

Small holes in your cervical esophagus may heal on their own without surgery. This is more likely to occur if fluid flows back into the esophagus and doesn't leak into your chest. You should determine if you need surgery within a day of your diagnosis. Most people with a perforated esophagus do need surgery. This is especially true if the hole is located in your chest or abdomen. During the procedure, your surgeon will remove scar tissue from the area around the perforation and then sew the hole shut. Very large perforations may require the removal of a portion of the esophagus. This procedure is called a partial esophagectomy. After the piece is removed, the remaining section of the esophagus is reconnected to the stomach.

Early treatment will include draining any fluid from your chest. You'll also need to take antibiotics to prevent or treat an infection. You won't be allowed to eat or drink anything by mouth until your treatment is completed. Your doctor will give you antibiotics and fluids intravenously, or through an IV. You may get nutrients through a feeding tube.

Subcutaneous Emphysema

Subcutaneous emphysema is when gas or air is in the layer under the skin. Subcutaneous refers to the tissue beneath the skin, and emphysema refers to trapped air (figure 24). It is sometimes abbreviated (SCE) or (SE) and also called tissue emphysema. Since the air generally comes from the chest cavity, subcutaneous emphysema usually occurs on the chest, neck and face, where it is able to travel from the chest cavity along the fascia. Subcutaneous emphysema has a characteristic crackling feel to the touch (figure 25), a sensation that has been described as similar to touching Rice Krispies; this sensation of air under the skin is known as subcutaneous crepitation.

Figure (24): Subcutaneous emphysema – Figure (25) crackling like touch

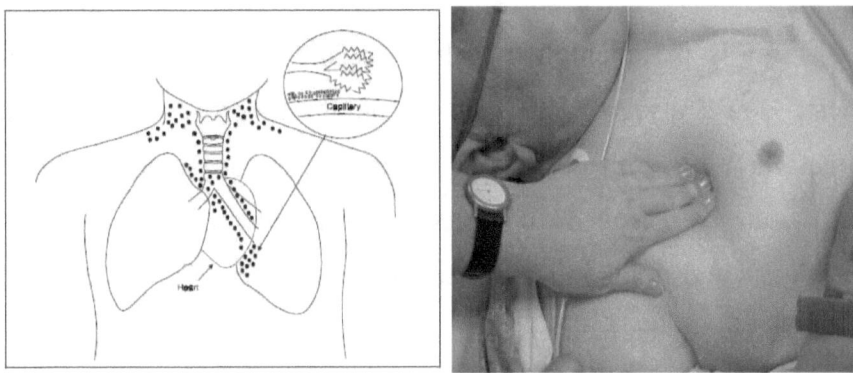

Causes:

Conditions that cause subcutaneous emphysema may result from both blunt and penetrating trauma; (SCE) is often the result of a stabbing or gunshot wound. Subcutaneous emphysema is often found in car accident victims because of the force of the crash.

Chest trauma, a major cause of subcutaneous emphysema, can cause air to enter the skin of the chest wall from the neck or lung. When the pleural membranes are punctured, as occurs in penetrating trauma of the chest, air may travel from the lung to the muscles and subcutaneous tissue of the chest wall. When the alveoli of the lung are ruptured, as occurs in pulmonary laceration, air may travel beneath the visceral pleura (the membrane lining the lung), to the hilum of the lung, up to the trachea, to the neck and then to the chest wall. The condition may also occur when a fractured rib punctures a lung; in fact, (27%) of patients who have rib fractures also have subcutaneous emphysema. Rib fractures may tear the parietal pleura, the membrane lining the inside of chest wall, allowing air to escape into the subcutaneous tissues.

Pathphysiology

Air is able to travel to the soft tissues of the neck from the mediastinum and the retroperitoneum (the space behind the abdominal cavity) because these areas are connected by fascial planes. From the punctured lungs or airways, the air travels up the perivascular sheaths and into the mediastinum, from which it can enter the subcutaneous tissues. Spontaneous subcutaneous emphysema is thought to result from increased pressures in the lung that cause alveoli to rupture. In spontaneous subcutaneous emphysema, air travels from the ruptured alveoli into the interstitium and along the blood vessels of the lung, into the mediastinum and from there into the tissues of the neck or head

Signs and symptoms of spontaneous subcutaneous emphysema vary based on the cause, but it is often associated with swelling of the neck and chest pain, and may also involve sore throat, neck pain, difficulty swallowing, wheezing and difficulty breathing. Chest X-rays may show air in the mediastinum, the middle of the chest cavity. A significant case of subcutaneous emphysema is easy to detect by touching the overlying skin; it feels like tissue paper or Rice Krispies. Touching the bubbles causes them to move and sometimes make a crackling noise.

The air bubbles, which are painless and feel like small nodules to the touch, may burst when the skin above them is palpated. The tissues surrounding SCE are usually swollen. When large amounts of air leak into the tissues, the face can swell considerably. In cases of subcutaneous emphysema around the neck, there may be a feeling of fullness in the neck, and the sound of the voice may change. If SCE is particularly extreme around the neck and chest, the swelling can interfere with breathing. The air can travel to many parts of the body, including the abdomen and limbs, because there are no separations in the fatty tissue in the skin to prevent the air from moving.

Diagnostic procedures:

Significant cases of subcutaneous emphysema are easy to diagnose because of the characteristic signs of the condition. In some cases, the signs are subtle, making diagnosis more difficult. Medical imaging is used to diagnose the condition or confirm a diagnosis made using clinical signs. On a chest radiograph, subcutaneous emphysema may be seen as radiolucent striations in the pattern expected from the pectoralis major muscle group. Air in the subcutaneous tissues may interfere with radiography of the chest, potentially obscuring serious conditions such as pneumothorax. It can also and reduce the effectiveness of chest ultrasound. On the other hand, since subcutaneous emphysema may become apparent in chest X-rays before a pneumothorax does, its presence may be used to infer that of the latter injury. Subcutaneous emphysema can also be seen in CT scans, with the air pockets appearing as dark areas (figure 26). CT scanning is so sensitive that it commonly makes it possible to find the exact spot from which air is entering the soft tissues.

Figure (26): Bubbles of air in the subcutaneous tissue (arrow) feel like mobile nodules

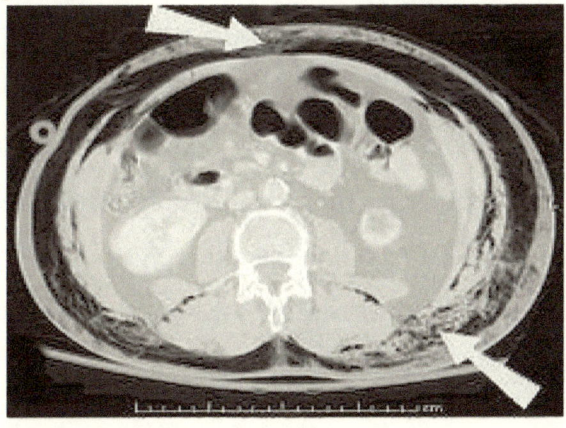

Management:

Most of the time, (SCE) itself does not need treatment however, if the amount of air is large, it can interfere with breathing and be uncomfortable. When the amount of air pushed out of the airways or lung becomes massive, usually due to positive pressure ventilation, the eyelids swell so much that the patient cannot see. The latter are urgent situations requiring rapid, adequate decompression. Severe cases can compress the trachea and do require treatment.

In severe cases of subcutaneous emphysema, catheters can be placed in the subcutaneous tissue to release the air. Small cuts, or "blow holes", may be made in the skin to release the gas. When subcutaneous emphysema occurs due to pneumothorax, a chest tube is frequently used to control the latter; this eliminates the source of the air entering the subcutaneous space. Suction may also be applied to the tube to remove air faster. The progression of the condition can be monitored by marking the boundaries with a special pencil for marking on skin.

Since treatment usually involves dealing with the underlying condition, cases of spontaneous subcutaneous emphysema may require nothing more than bed rest, medication to control pain, and perhaps supplemental oxygen. Breathing oxygen may help the body to absorb the subcutaneous air more quickly. Reassurance and observation are also part of treatment.

Nursing Management of chest traumatic patients:

Levels of chest trauma Prevention:

1- Primary prevention (prevent the event**).**

☐ Drivingsafety classes

☐ Speedlimits

☐ Campaignsto not drink and drive

2- Secondary prevention (minimize the impact of the traumatic event).

☐ Seatbelt use

☐ Airbags

☐ Carseats

☐ Helmets

3- Tertiary prevention

maximize patient outcomes after a traumatic event through emerge ncy response systems, medical care, and rehabilitation

Pre-hospital Care (save Transport)

☐ Emergency stabilization and quick transport

☐ ABCs(with cervical stabilization)

☐ IVaccess and fluid administration

☐ Controlhemorrhaging

☐ Stabilizefractures

Triaging the chest traumatic patients (through primary and secondary survey):

Primary Survey

☐ Done in 1 to 2 minutes

☐ Airway patency (with C-spine immobile)

☐ Breathing effectiveness

☐ Circulation, including hemorrhage and pulses

☐ Disability (overview of neurological status)

☐ Identify life-threatening injuries accurately to establish priorities

Secondary Survey

☐ Performed after life-threatening injuries are identified and treated

☐ Examination of all body systems: head-to-toe and front-to-back

☐ Maintain C-spine immobilization until cleared by x-ray

☐ X-ray studies (as determined

Common Nursing problems and how to manage:

1- Maintain Airway Patency

☐ Many factors affect the airway (e.g., facial fractures, bleeding, v omiting, and decreased sensorium)

☐ Nasopharyngeal airways: used in spontaneously breathing patient.

☐ Endotracheal intubation often required injury)

☐ Laboratory studies

2- Ineffective Breathing

☐ Ongoing assessment is essential

☐ Respiratory status

☐ Arterial blood gases (ABG)

- ☐ Chestx-rays
- ☐ Computedtomography (CT) imaging
- ☐ Improveventilation and gas exchange
- ☐ Mechanicalventilation
- ☐ Needlethoracostomy and chest tube insertion
- ☐ Administrationof fluids and blood products
- ☐ Administrationof sedation and analgesics

3- Impaired Gas Exchange

- ☐ Causes
- ☐ Decreasein inspired air
- ☐ Retainedsecretions
- ☐ Lungcollapse or compressed
- ☐ Atelectasis

Nursing Approach to the Client with Chest truama, Nursing Assessment:

1. Assess for history of the injury.

2. Assess presence of signs and symptoms of impaired respiratory function (dyspnea, chest pain, asymmetric chest movements, signs of paradoxical breathing, cyanosis, anxiety, bloody sputum)

3. Assess chest wall for presence of wounds and fractures.

4. Assess signs of increased intrathoracic pressure (mediastinal shift, trachea shift, progressive signs of respiratory and cardiovascular insufficiency).

5. Lung auscultation shows diminution or absence of breathing sounds on the affected side.

6. Assess vital signs, CVP, ECG, fluid balance.

7. Assess diagnostic tests and procedures for abnormal values (chest x-ray, CT, pleural puncture).

Nursing Diagnosis:

1. Increased risk of hypoxia and respiratory failure related to injury.

2. Increased risk of hypovolemia and shock related to hemorrhage and impaired cardiac function.

3. Pain related to injury.

4. Anxiety related to the symptoms of disease and fear of death.

Nursing Plan and Interventions Goals:

1. Maintain respiratory and cardiovascular function.

2. Prevent avoidable injury and complications.

3. Then surgical intervention prescribed, prevent postoperative complications.

4. Relief or diminish symptoms.

5. Decreased anxiety with increased knowledge.

Interventions:

1. Assess, report , and record signs and symptoms and reactions to treatment.

2. Observe respiratory status closely, report immediately if changed.

3. Monitor vital signs, fluid balance, level of consciousness closely.

4. Administer oxygen and other medications as prescribed, monitor for side effects.

5. Maintain patency of chest tubes, observe appropriate connections and presence of negative pressure in system.

6. Administer blood transfusions and IV therapy as prescribed, monitor for side effects.

7. Place client in the high-Fowler position then has chest injury, on a side of the chest tube insertion then hemothorax presents to provide drainage.

8. Monitor laboratory tests results for abnormal values.

9. Prepare client and his family for surgical intervention.

10. For client after surgical intervention provide postoperative care and observe possible postoperative complications.

11. Encourage the client to turn and cough and breath deeply.

12. Observe signs of possible secondary pulmonary infection, report immediately.

13. Provide appropriate skin care to prevent pressure sores.

14. Provide emotional support to client, explain all procedures to decrease anxiety and to obtain cooperation.

15. Instruct client regarding disease, diagnostic procedures, treatment and its complications, home care, daily activities, restrictions and follow-up.

Evaluation:

1. Maintain adequate respiratory function and gas exchange.

2. No evidence of complications.

3. Maintains stable vital signs, fluid balance, and nutritional state.

4. Laboratory tests results shows no abnormalities.

Notes

References:

American Nurses Association   code of ethics (O'Neill, 2003).

Arras, J. (1993). Ethical issues in emergency care. Clinics in geriatric medicine, 9(3), 655-664.

Billeter, Adrian T.; Druen, Devin; Franklin, Glen A.; Smith, Jason W.; Wrightson, William; Richardson, J. David (2013). "Video-assisted thoracoscopy as an important tool for trauma surgeons: a systematic review". Langenbeck's Archives of Surgery 398 (4): 515–523. doi:10.1007/s00423-012-1016-7. ISSN 1435-2443.

Brun-Buisson, C., Minelli, C., Bertolini, G., Brazzi, L., Pimentel, J., Lewandowski, K., ... & ALIVE Study Group. (2004). Epidemiology and outcome of acute lung injury in European intensive care units. Intensive care medicine, 30(1), 51-61.

Calhoon, J. H., & Trinkle, J. K. (1997). Pathophysiology of chest trauma. Chest surgery clinics of North America, 7(2), 199-211.

Casey, R. & Emde, K. (2008). Displaced fractured sternum following blunt chest trauma. Journal of Emergency Nursing. 34(1), 83-85.

Clancy, K., Velopulos, C., Bilaniuk, J. W., Collier, B., Crowley, W., Kurek, S., Lui, F., Nayduch, D., Sangosanya, A., Tucker, B. & Haut, E.R. (2012). Screening for blunt cardiac injury: An Eastern Association for the Surgery of Trauma practice management guideline. Journal of Trauma and Acute Care Surgery, 73, S301-S306.

Collins, J. (2000). Chest wall trauma. Journal of Thoracic Imaging, 15(2), 112–119.

Dalal, S., Nityasha, V. M., & Dahiya, R. S. (2009). Prevalence of chest trauma at an apex institute of North India: A retrospective study. Internet J Surg, 18, 1.

Demirhan, R., Onan, B., Oz, K., & Halezeroglu, S. (2009). Comprehensive analysis of 4205 patients with chest trauma: a 10-year experience. Interactive cardiovascular and thoracic surgery, 9(3), 450-453.

Flynn, M. B., & Bonini, S. (1999) Blunt chest trauma: Case report. Critical Care Nurse, 19(5), 68–77.

 Huggins JT, Sahn SA (2004). "Causes and management of pleural fibrosis". Respirology 9 (4): 441–7. doi:10.1111/j.1440-1843.2004.00630.x. PMID 15612954.

Karlet, M. C. (1997). Update for nurse anesthetists: Thoracic trauma. American Association of Nurse Anesthetists Journal, 65(1), 73–80.

Kramer, J. L. (2002). Pathophysiology of Thoracic Trauma. In Seminars in Cardiothoracic and Vascular Anesthesia (Vol. 6, No. 2, pp. 57-61). SAGE Publications.

Kulshrestha, P., Munshi, I., & Wait, R. (2004). Profile of chest trauma in a level I trauma center. Journal of Trauma and Acute Care Surgery, 57(3), 576-581.

Light RW (2010). "Pleural effusion in pulmonary embolism". Semin Respir Crit Care Med 31 (6): 716–22. doi:10.1055/s-0030-1269832. PMID 21213203.

LoCicero 3rd, J., & Mattox, K. L. (1989). Epidemiology of chest trauma. The Surgical clinics of North America, 69(1), 15-19.

McElroy S.,(2012)
http://nursing.advanceweb.com/Features/Articles/Ethical-Considerations-in-Emergency-Nursing.aspx

nursingapproachchestinjuries.blogspot.com/

Pape, H. C., Remmers, D., Rice, J., Ebisch, M., Krettek, C., & Tscherne, H. (2000). Appraisal of early evaluation of blunt chest trauma: Development of a standardized scoring system for initial clinical decision making. Journal of Trauma-InjuryInfection & Critical Care, 49(3), 496–504.

Reay, G., & Rankin, J. A. (2013). The application of theory to triage decision-making. International emergency nursing, 21(2), 97-102.

Rousset, P.; Rousset-Jablonski, C.; Alifano, M.; Mansuet-Lupo, A.; Buy, J.-N.; Revel, M.-P. (2014). "Thoracic endometriosis syndrome: CT and MRI features". Clinical Radiology 69 (3): 323–330. doi:10.1016/j.crad.2013.10.014. ISSN 0009-9260.

Weldon, Erin; Williams, Jen (2012). "Pleural Disease in the Emergency Department". Emergency Medicine Clinics of North America 30 (2): 475–499.doi:10.1016/j.emc.2011.10.012. ISSN 0733-8627.

WHO, (2012); Annual injury surveillance report Egypt, available at:www.emro.who.int/dsaf/dsa1087.pdf

Wolf, L., Brysiewicz, P., LoBue, N., Heyns, T., Bell, S. A., Coetzee, I., ... & Hangula, R. (2012). Developing a framework for emergency nursing practice in Africa. African Journal of Emergency Medicine, 2(4), 174-181.

Yahia, A., Ali, N. S., & Elhabashy, S. (2013). Factors Affecting Validity of Arterial Blood Gases Results among Critically Ill Patients: Nursing Perspectives. *Journal of Education and Practice*, *4*(15), 43-56.

www.ingramcontent.com/pod-product-compliance
Lightning Source LLC
Chambersburg PA
CBHW022111170526
45157CB00004B/1585